Specialist Periodical Reports Online

Application for Free Access to Electronic Chapters

Catalysis
Volume 18

Customers purchasing a print volume are now entitled to free site-wide access to the electronic version of that title, including back volumes from 1998 - 2002 inclusive.
(For the definition of a site, please consult **www.rsc.org/subagree**)

To apply for free access, please complete and return this form.

Contact Name: ____________________

Job Title: ____________________

Organisation: ____________________

Address: ____________________

____

____

Town: ____________________

Country: ____________________

Post/Zip Code: ____________________

E-mail: ____________________

Telephone: ____________________

RS•C

Continued over

– Side 2 –

Please give IP addresses of the site overleaf in the format w.x.y.z.

Wildcards and ranges are allowed

e.g. for Class B 128.128.*.*

for Class C 192.192.192.*

for a range 123.456.1-99.*

Please put each entry on a new line:

____

____

____

____

(please use a continuation page if necessary)

I have read and agree to the terms of the Electronic Information Licence Agreement at **www.rsc.org/subagree**

Signed:____________________ **Date**________________

☐ Please contact me about access for additional sites outside the site definition in the Electronic Information Licence Agreement

☐ Please arrange username/password access and not IP address control

Please return this form by non-electronic means to the Royal Society of Chemistry at the address below.

Photocopies or facsimiles are not acceptable.

Sales & Customer Care Dept

Royal Society of Chemistry · Thomas Graham House

Science Park · Milton Road · Cambridge · CB4 0WF · UK

T +44(0)1223 432360 · F + 44(0)1223 426017 · E sales@rsc.org

Registered Charity No. 207890

The Royal Society of Chemistry will store the information you supply on its electronic records in order that information about its activities, products and services may be sent to you by mail, telephone, email or fax.

If you **DO NOT** wish to receive information, please put a tick in the box ☐.

Catalyis

Volume 18

A Specialist Periodical Report

Catalysis

Volume 18

A Review of Recent Literature

Senior Reporter
J.J. Spivey, *Louisiana State University, Baton Rouge, LA, USA*

Reporters
Moises A. Carreon, *Michoacan, Mexico*
Gabriele Centi, *University of Messina, Italy*
Steven S. C. Chuang, *The University of Akron, OH, USA*
Vadim V. Guliants, *University of Cincinnati, OH, USA*
Moon Hyeon Kim, *Daegu University, Korea*
Siglinda Perathoner, *University of Messina, Italy*
In-Sik Nam, *Pohang University of Science and Technology (POSTECH), Korea*
Mark G. White, *Georgia Institute of Technology, GA, USA*

RS•C
advancing the chemical sciences

If you buy this title on standing order, you will be given FREE access to the chapters online. Please contact sales@rsc.org with proof of purchase to arrange access to be set up.

Thank you.

ISBN 0-85404-234-2
ISSN 0140-0568

A catalogue record for this book is available from British Library

Published by The Royal Society of Chemistry,
Thomas Graham House, Science Park, Milton Road, Cambridge CB4 0WF, UK

Registered Charity Number 207890

For further information see our web site at www.rsc.org

Typeset by Vision Typesetting, Manchester, UK
Printed by Athenaeum Press Ltd, Gateshead, Tyne & Wear

Preface

The application of catalytic principles to energy and chemical production as well as the challenges of environmental issues remains an important field of study. There continue to be new ideas, new problems to be solved, and new insight into how catalysts work. It is my hope that this volume of the Catalysis book series will be a resource for those who are working in this exciting area.

This volume consists of reviews devoted to a range of important subjects. Vadim Guliants and Moises Carreon (University of Cincinnati) review the selective oxidation of butane. This is an excellent example of a catalytic process designed to add value to an inexpensive raw material, and is the only vapor phase selective oxidation of an alkane that is practiced industrially. This process also avoids the use of benzene, which eliminates the risk of handling this carcinogenic compound. The authors review the synthesis, activation, and mechanism of this reaction on V-P-O catalysts.

Gabriele Centi and Siglinda Perathoner (University of Messina, Italy) examine the use of solid catalysts for the removal of contaminants from water supplies. This includes photocatalytic processes as well as oxidation and reduction reactions. There are a wide range of catalysts used in these various processes. In addition to their activity, deactivation is often a critical concern. The authors show that there are significant challenges remaining in this area.

The use of polynuclear metal complexes provides a novel approach to the synthesis of supported metals and metal oxides. Mark White (Georgia Tech) shows how these materials can be synthesized and characterized. Quantum mechanics are used to compare predicted and experimental results of the interactions of the metal complexes and surfaces. The use of these materials as both adsorbents and catalysts suggests the importance of understanding both how they are synthesized, and their post-synthesis structure.

In-Sik Nam and Moon Hyeon Kim (Pohang University of Science and Technology, Korea) review new materials for the selective catalytic reduction of NO_x from combustion processes. Despite significant research efforts over the last 20 years, there are still unresolved issues, such as inhibition and deactivation by steam. The authors show how new synthesis methods, especially for zeolites, can be used to improve these catalysts.

Finally, Steve Chuang (University of Akron) reviews catalysts for use in solid oxide fuel cells. These fuel cells hold the promise of significant improvements in energy efficiency with minimum emissions, but challenges remain – particularly for the catalysts. A particular challenge is the effect of the high temperature operation on the oxide electrolyte. Also important is the stability of the anode in the reducing environment and its tolerance to sulfur. These and other constraints on the catalysts used in this fuel cell are the subject of this review.

I wish to thank the authors for the effort they have put into these chapters, and the Royal Society of Chemistry for their support, particularly Janet Freshwater. Comments and suggestions are welcome.

James J. Spivey
Department of Chemical Engineering
Louisiana State University
Baton Rouge, LA 70803
jjspivey@lsu.edu

Contents

Catalysis, Volume 18

1

Vanadium-Phosphorus-Oxides: from Fundamentals of *n*-Butane Oxidation to Synthesis of New Phases

BY VADIM V. GULIANTS AND MOISES A. CARREON

1 Introduction

The abundance and low cost of light alkanes have generated in recent years considerable interest in their oxidative catalytic conversion to olefins, oxygenates and nitriles in the petroleum and petrochemical industries [1-4]. Rough estimates place the annual worth of products that have undergone a catalytic oxidation step at \$20-40 billion worldwide [4]. Among these, the 14-electron selective oxidation of *n*-butane to maleic anhydride (2,5-furandione) on vanadium-phosphorus-oxide (VPO) catalysts is one of the most fascinating and unique catalytic processes [4,5]:

It is the only industrial process of a selective vapor-phase oxidation of an alkane that uses dioxygen [5]. The demand for maleic anhydride comes principally from the manufacture of unsaturated polyester resins, agricultural chemicals, food additives, lubricating oil additives, and pharmaceuticals [6].

Bergman and Frisch [7] disclosed in 1966 that selective oxidation of *n*-butane was catalyzed by the VPO catalysts, and since 1974 *n*-butane has been increasingly used instead of benzene as the raw material for maleic anhydride production due to lower price, high availability in many regions and low environmental impact [8]. At present more than 70 % of maleic anhydride is produced from *n*-butane [6]. However, productivity from *n*-butane is lower than in the case of benzene due to lower selectivities to maleic anhydride at higher conversions and somewhat lower feed concentrations ($<$ 2 mol. %) used to avoid flammability of a process stream. Under typical industrial conditions (2 mol. % *n*-butane in air, 673-723K, and space velocities of 1100-2600 h^{-1}) the selectivities [9] for fixed-bed production of maleic anhydride from *n*-butane are 67-75 mol. % at 70-85 % *n*-butane conversion [10]. Another unique feature of the VPO catalysts is that no support is used in partial oxidation of *n*-butane.Many studies of *n*-butane oxidation on the VPO catalysts indicated that crystalline vanadyl(IV) pyrophos-

Catalysis, Volume 18

phate, $(VO)_2P_2O_7$, is present in the most selective catalysts, e.g. [10-12]. However, the VPO system is characterized by facile formation and interconversion of a number of crystalline and amorphous V^{III}, V^{IV} and V^{V} phosphates [10]. Various research groups detected these phases in the VPO catalysts and proposed different models of the active and selective VPO phase and surface sites in *n*-butane oxidation [10-13].

The VPO catalysts are prepared by thermal dehydration of its precursor, vanadyl(IV) hydrogen phosphate hemihydrate, $VOHPO_4{\cdot}0.5H_2O$. The catalytic performance of the VPO catalysts depends on (i) the method of $VOHPO_4{\cdot}0.5H_2O$ synthesis (types and concentrations of reagents, reducing agents and solvents, the reduction temperature and synthesis duration), (ii) the procedures for activation and conditioning of the precursor at high temperature and (iii) the nature of metal promoters. These factors important for understanding the catalytic behavior of the VPO system in *n*-butane oxidation were discussed previously in a number of excellent early reviews [10-14]. Therefore, in this chapter we briefly go over the conclusions of early studies and discuss in greater detail recent findings with emphasis on fundamental aspects of VPO catalysis, such as the mechanism of $VOHPO_4{\cdot}0.5H_2O$ formation and its transformation to active and selective VPO catalysts, the mechanism of *n*-butane oxidation, the role of promoters and the synthesis of new VPO phases. It is expected that new fundamental insights into molecular structure and catalytic function of this unique catalytic system will lead to the design of improved mixed metal oxide catalysts for selective oxidation of light alkanes.

2 Synthesis of $VOHPO_4{\cdot}0.5H_2O$ Precursor

There is a general agreement in the VPO literature [4, 10, 14-21] that the necessary synthesis conditions to obtain an optimal catalyst are the following: (i) synthesis of microcrystalline $VOHPO_4{\cdot}0.5H_2O$ in an alcohol characterized by the preferential exposure of the basal (001) planes, (ii) the presence of defects in the stacking of the (001) planes and (iii) a slight excess of phosphate with respect to the stoichiometric amount employed in the synthesis (P/V = 1.01-1.10). This excess phosphate is strongly bound to the surface and cannot be removed by simple washing of the precursor in polar solvents.

Three major synthesis methods were reported for preparation of the $VOHPO_4{\cdot}0.5H_2O$ precursor:

1. In *aqueous synthesis*, V^{V} compounds (e.g. V_2O_5) are reduced to V^{IV} in aqueous solutions of orthophosphoric acid, followed by evaporation of the solvent to dryness [22]:

$$V_2O_5 + 2NH_2OH\text{-}HCl + 2H_3PO_4 \rightarrow 2VOHPO_4{\cdot}0.5H_2O + N_2 + 2HCl + 4H_2O$$

2. In *organic synthesis*, V^{V} compounds are reduced by an anhydrous alcohol, followed by the addition of anhydrous orthophosphoric acid dissolved in the same alcohol and precipitation of $VOHPO_4{\cdot}0.5H_2O$ [16, 19]:

$$V_2O_5 + EtCHOHMe + 2H_3PO_4 \rightarrow 2VOHPO_4{\cdot}0.5H_2O + EtCOMe + 2H_2O$$

3. In *model organic synthesis*, V^V orthophosphate dihydrate, $VOPO_4{\cdot}2H_2O$, is first synthesized from V_2O_5 and H_3PO_4 in aqueous medium and then reduced to $VOHPO_4{\cdot}0.5H_2O$ by an alcohol in a separate step:

$$V_2O_5 + 2H_3PO_4 + H_2O \rightarrow 2VOPO_4{\cdot}2H_2O + H_2O$$

$$2VOPO_4{\cdot}2H_2O + EtCHOHMe \rightarrow 2VOHPO_4{\cdot}0.5H_2O + EtCOMe + 3H_2O$$

The *organic synthesis* usually provides the most active and selective catalysts [4,10,14,16,19]. All three methods may also lead to various hydrated vanadyl(IV) hydrogen phosphate phases, $VOHPO_4{\cdot}nH_2O$ ($n = 0.5$, 1, 2, 3, and 4), which are all precursors of the VPO catalysts. The precursor with $n = 0.5$ ($VOHPO_4{\cdot}0.5H_2O$) produces the best VPO catalysts [10]. Another phase, $VO(H_2PO_4)_2$, is observed when a considerable excess of phosphate is used in the organic synthesis ($P/V > 2$) [16]. The main differences observed in the VPO precursors obtained by various methods is the morphology of the $VOHPO_4{\cdot}0.5H_2O$ crystallites. The XRD patterns of $VOHPO_4{\cdot}0.5H_2O$ synthesized by aqueous and organic methods and, accordingly, referred to as *organic* and *aqueous* VPO precursors and catalysts are shown in Figure 1. These patterns indicate that organic precursors are less crystalline and preferentially expose the (001) planes [14], as manifested in a broader (001) reflection and its lower relative intensity as com-

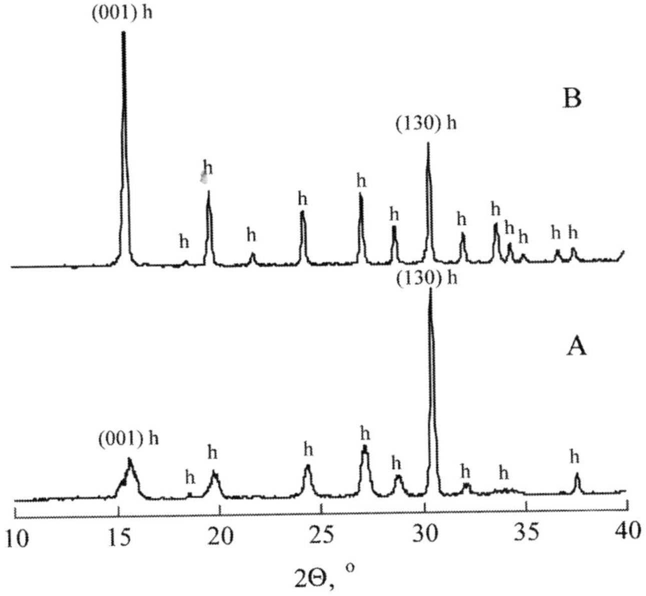

Figure 1 *XRD patterns of the organic (A) and aqueous (B) precursors of VPO catalysts [63]. h= $VOHPO_4{\cdot}0.5H_2O$. Peak width: FWHM of (001)= 0.56 and 0.18°; FWHM of (130)= 0.18 and 0.22° in organic and aqueous precursors, respectively. Relative peak intensity: I(001)/I(130)= 0.20 and 1.95 in organic and aqueous precursors, respectively*

pared to the intensity of the in-plane (130) reflection (Figure 1).

The morphology of organic precursors depends on many factors, e.g. (i) the nature of the solvent/reducing agent (an aliphatic or benzylic alcohol) [16, 19], (ii) the synthesis P/V ratio [16], (iii) the time and temperature of reduction [19], and (iv) the amount of water present during synthesis [16].

Currently only organic VPO catalysts are employed industrially in *n*-butane oxidation as the most active and selective. The following steps have been proposed in the formation of the $VOHPO_4{\bullet}0.5H_2O$ in organic medium [19]: (i) the formation of colloidal V_2O_5 at the water-alcohol interface, (ii) the dissolution of V_2O_5 through the formation of V^V-alcoholate species, (iii) the reduction of the dissolved V^V-alcoholate species in the liquid phase to solid V_2O_4, and (iv) the reaction of V_2O_4 with H_3PO_4 to form $VOHPO_4{\bullet}0.5H_2O$ at the solid-liquid interface. The type of aliphatic alcohol influences the temperature of reduction of V^V which is kinetically controlled and complete only at long reduction times upon addition of benzyl alcohol and orthophosphoric acid [19].

In the reduction by benzyl alcohol, many studies reported the formation of $VOHPO_4{\bullet}0.5H_2O$ platelets possessing stacking faults of the (001) planes seen in the preferential broadening of the (001) reflection. The stacking faults develop due to the trapping of alcohol molecules between the (001) layers of the precursor and their release during precursor transformation to $(VO)_2P_2O_7$ [4, 10, 16, 19]. The effect of the above synthesis parameters on the properties of $VOHPO_4{\bullet}0.5H_2O$ is to vary the exposure of the (001) plane, create the stacking fault strain in the crystallites and influence the degree of V^V reduction.

In both aqueous and organic syntheses, the $VOHPO_4{\bullet}0.5H_2O$ precursor has a P/V ratio higher than the stoichiometric value [10, 16, 19]. The maximum value corresponds to $P/V = 1.1$, while the excess phosphate remains in synthesis solution. The X-ray photoelectron (XPS) analysis indicates that the excess phosphate is localized at the surface of the vanadyl pyrophosphate catalysts (surface $P/V = 1.5$-3.0) [23].

The well-known redox chemistry of V^V provides important insights into the mechanism of the $VOHPO_4{\bullet}0.5H_2O$ precursor formation in organic medium. Waters and Littler [24] have shown that most V^V reductions proceed via a free-radical mechanism where complexation of V^V to alcohol precedes the one-electron transfer step, i.e. an inner sphere electron transfer. Waters and Littler [24] proposed ternary tetrahedral complex formation between VO_2^+, H_3O^+ and R_2CHOH to yield $[V(OH)_3OHCHR_2]^{2+}$ and observed the following kinetic expression

$$-d[V^V]/dt = k[R_2CHOH][VO_2^+][H_3O^+]$$

corresponding to slow decomposition of this species to V^{IV} and an alcohol radical, $R_2C^{\bullet}$-OH. The formation of the protonated complex should be assisted by the more acidic medium which was observed experimentally in numerous studies [24-26]. The inertness of simple tertiary alcohols toward V^V indicated that the α-C-H bond is involved in the reaction which was confirmed by measuring the primary isotope effect produced by deuterium substitution at this position in cyclohexanol [24]. A mechanism involving cyclic transfer of the α-H atom to a coordination sphere of the metal ion has been proposed:

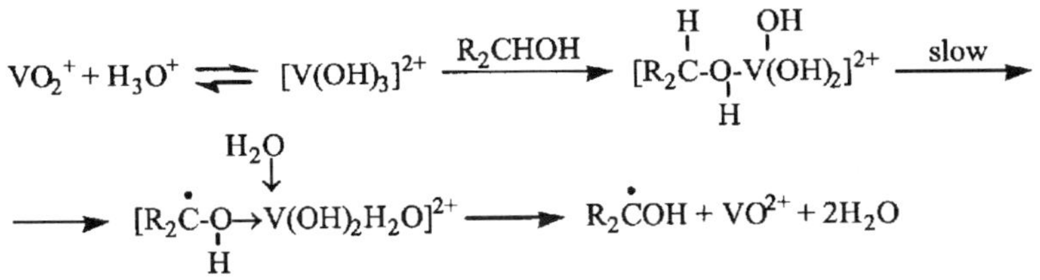

This mechanism shows that the V^V reduction during the VPO precursor synthesis is slow at low H_3O^+ concentration. The rate of reduction is slow even in the presence of anhydrous H_3PO_4, since $[H_3O^+]$ is low in the absence of water. The reaction is likely autocatalytic: the water evolved during the reduction is protonated and the resultant H_3O^+ accelerates the reaction rate.

Rocek and Aylward [27] used substituted cyclobutanols as a probe into the mechanism of V^V reduction by alcohols. Oxidation of cyclobutanols is a widely accepted method for discerning one- and two-electron processes, as they lead to cyclobutanones and γ-hydroxyaldehydes, respectively [28-30]. Rocek and Aylward [27] observed the formation of γ-hydroxyaldehydes in V^V reductions providing evidence of a one-electron process. The V^V oxidation of cyclobutanols is first order in both the cyclobutanols and the protonated monomeric HVO_2^+ (aq.) species. HVO_2^+ (aq.) is a stronger oxidant than VO_2^{2+} (aq.), and at the acidic conditions employed in these oxidations, V^V was predominantly in its monomeric form (HVO_2^+), despite its pronounced tendency toward the formation of polymeric ions [31]. Methyl cyclobutyl ether was found to be 10^4 times less reactive than cyclobutanol. This striking difference in reactivity indicated that the OH bond plays a vital role in the oxidation process and is broken either prior to or during the rate-determining step, as explained by formation of an ester of vanadic acid intermediate suggested as $ROV(OH)_2(OH_2)_n^{2+}$ ($n=1$, 2, or 3).

Tracey and Gresser [32] confirmed spontaneous formation of vanadate esters in aqueous solutions of alcohols in ^{51}V NMR spectra (pH=7-11, 1-50 mM, 328K) where the vanadate occurs as the tetrahedral mono- or diprotonated anions, HVO_4^{2-} and $H_2VO_4^-$ denoted as V_i, dimeric ($HO_3VOVO_3H^{2-}$ or V_2), and tetrameric ($H_4V_4O_{14}^{4-}$ or V_4) species. The complexity of esterification of vanadate arises in part from the ability of vanadate (VO_4^{3-}) to undergo protonation and oligomerization as pH and concentration are changed [31,33,34]. Tracey and Gresser showed that vanadate complexes with monodentate hydroxylic ligands are tetrahedral analogs of phosphate esters. Equilibrium constants for the formation of monoanionic alkyl vanadate esters from V_i monoanion and alcohols, $K_f=[ROVO_3H^-]/([H_2VO_4^-][ROH])$, are about 0.2 M^{-1}, and relatively insensitive to the pK_a of the alcohol and to whether the alcohol is primary, secondary, or tertiary [35-38]. Since the pK_a values of V_i and alkylvanadates are above 8.0, this K_f value can be used to estimate the concentration of a given vanadate ester in solution at a neutral pH containing known concentrations of V_i and alcohol. In studies of V_i in aqueous methanol Tracey and co-workers [38] found that methyl esters of *divanadate* can also form spontaneously with equilibrium constants similar to those for formation of esters of monovanadate: V_2 + MeOH $\rightarrow V_2(OMe)$ + H_2O, $K = 3.0\ M^{-1}$.

Gresser *et al.* [39] observed formation of mixed anhydrides of vanadate with phosphate and pyrophosphate in aqueous solutions by ^{51}V NMR, which may be considered as molecular precursors of vanadium phosphate phases. The formation of the mixed anhydrides was in fact more favorable than that of phosphate anhydrides (i.e. pyrophosphate) by more than 10^6-fold in the K_f for the phosphovanadate as compared to pyrophosphate at neutral pH. Formation of the divanadate species (V_2) was 10^8 times more favored over formation of pyrophosphate. The reasons for the preferred formation of phosphovanadates may have to do with an interaction of the lone pairs of electrons on the bridging oxygen atom with orbitals on the adjacent vanadium and phosphorus atoms. If this occurred, it would cause the bridging oxygen to be more *sp*-like than sp^3, with the result that the P-O-V or V-O-V bond angle in the anhydride would be larger than the P-O-P angle in the pyrophosphate. If the V-O-V and P-O-V angles are large, then the corresponding anhydrides in contrast to the pyrophosphate [40] may not be able to chelate Mg^{2+} which has been confirmed experimentally [41].

Some of the ways the vanadate esters and phosphovanadates [32-39] may be involved in the reduction of V^V during the $VOHPO_4 \cdot 0.5H_2O$ precursor formation are:

A. Redox decomposition of monoalkyl esters of monomeric vanadate via a mechanism similar to that of Waters and Littler [24]. In this case, free radical species will be generated which can be detected by an ESR spin trapping technique.

B. Redox decomposition of monoalkyl ester of dimeric vanadate, $ROVO_2$-O-VO_3^{3-}, accompanied by two rapid successive one-electron transfers. Then, no free radical species are expected, and the products should be the carbonyl compound and VO^{2+}.This mechanism has not yet been established for vanadate oxidations.

C. Redox decomposition of monoalkyl esters of phosphovanadate, $[ROV^VO_2$-O-$PO_3H]^{2-}$, similar to case A. However, in this case the redox decomposition may lead to $[V^{IV}O(H_2O)_4OPO_3H]$ species which precedes the formation of $VOHPO_4 \cdot 0.5H_2O$ precursor. The $[V^{IV}O(H_2O)_4OPO_3H]$ species may be detected during early stages of V^V reduction by ESR.

3 Thermal Activation of $VOHPO_4 \cdot 0.5H_2O$ Precursor

A number of crystalline VPO phases were observed during the transformation of the $VOHPO_4 \cdot 0.5H_2O$ precursor to the active VPO catalyst [11, 12] depending on:

- the temperature, time, and atmosphere of activation;
- the morphology of the precursor;
- the P/V ratio in the precursor; and
- the presence defects in the structure.

Two different activation procedures are generally reported in the scientific and patent literature:

- Activation in an oxygen-free atmosphere at T > 673K, followed by introduction of the reactant mixture of *n*-butane in air. $VOHPO_4 \cdot 0.5H_2O$ transforms quantitatively to poorly crystalline $(VO)_2P_2O_7$ during the first step [16], which can be partially oxidized to V^V orthophosphates after the introduction of the reactant mixture [13].
- Calcination in air at T < 673K, after which the reactant mixture is introduced [15, 16, 23].

As the $VOHPO_4 \cdot 0.5H_2O$ precursor is heated, the trapped alcohol molecules are released, which creates structural defects, microcracks and increases the surface area. The precursor first transforms into an amorphous phase [15, 23], which can be further dehydrated to crystalline $(VO)_2P_2O_7$ and partially oxidized to V^V orthophosphates once the reactant mixture is introduced. Excess phosphate present in the precursor stabilizes $(VO)_2P_2O_7$ against overoxidation in oxygen-containing atmosphere [13]. The *in situ* Raman spectroscopy provided insights into the nature of the transformation of the model organic $VOHPO_4 \cdot 0.5H_2O$ precursor [42]. When the precursor phase was heated in an inert He atmosphere, a continuous loss of local order was observed up to 724K. At that temperature, fingerprint Raman bands of $VOHPO_4 \cdot 0.5H_2O$ were no longer visible, and instead weak features corresponding to poorly crystalline $(VO)_2P_2O_7$ emerged. The *in situ* spectra suggested that these two phases did not coexist during the precursor transformation in inert atmosphere. Similar observations were made during the precursor transformation in reactive *n*-butane-air environment. The precursor is completely transformed into a mixture of δ-$VOPO_4$ and $(VO)_2P_2O_7$ at 708K. The redox environment leads to partial oxidation of vanadyl(IV) species into a V(V) orthophosphate. The presence of V(V) orthophosphate is usually responsible for higher activity and lower selectivity of these catalysts as compared to the fresh catalysts activated in inert atmosphere. It is interesting that the transformation proceeds via formation of an amorphous VPO intermediate regardless of the gas environment, and that $VOHPO_4 \cdot 0.5H_2O$ and $(VO)_2P_2O_7$ phases are not observed simultaneously during the transformation. Figure 2 shows the different phase transformations in the VPO system [43]. Xue and Schrader further examined the VPO phase transformations using the *in situ* Raman spectroscopy [44]. They found that $(VO)_2P_2O_7$ experienced a structural disorder during prolonged exposure to water vapor at elevated temperatures which resulted in the V-O-P bond breaking, the loss of phosphate and the appearance of bulk vanadium oxide.

Electron microscopy studies [16, 19, 45] have shown that $(VO)_2P_2O_7$ maintains the morphology of the precursor. The XRD data are also consistent with the broadening of the interlayer spacing reflections in both $VOHPO_4 \cdot 0.5H_2O$ ((001) vs. the in-plane (130) reflection) and $(VO)_2P_2O_7$ ((200) vs. the in-plane (024) reflection) shown in Figure 1 and 3. Based on these observations as well as on close structural analogies of the two phases Bordes *et al.* [43] proposed that the transformation of $VOHPO_4 \cdot 0.5H_2O$ to $(VO)_2P_2O_7$ is topotactic, i.e. it proceeds

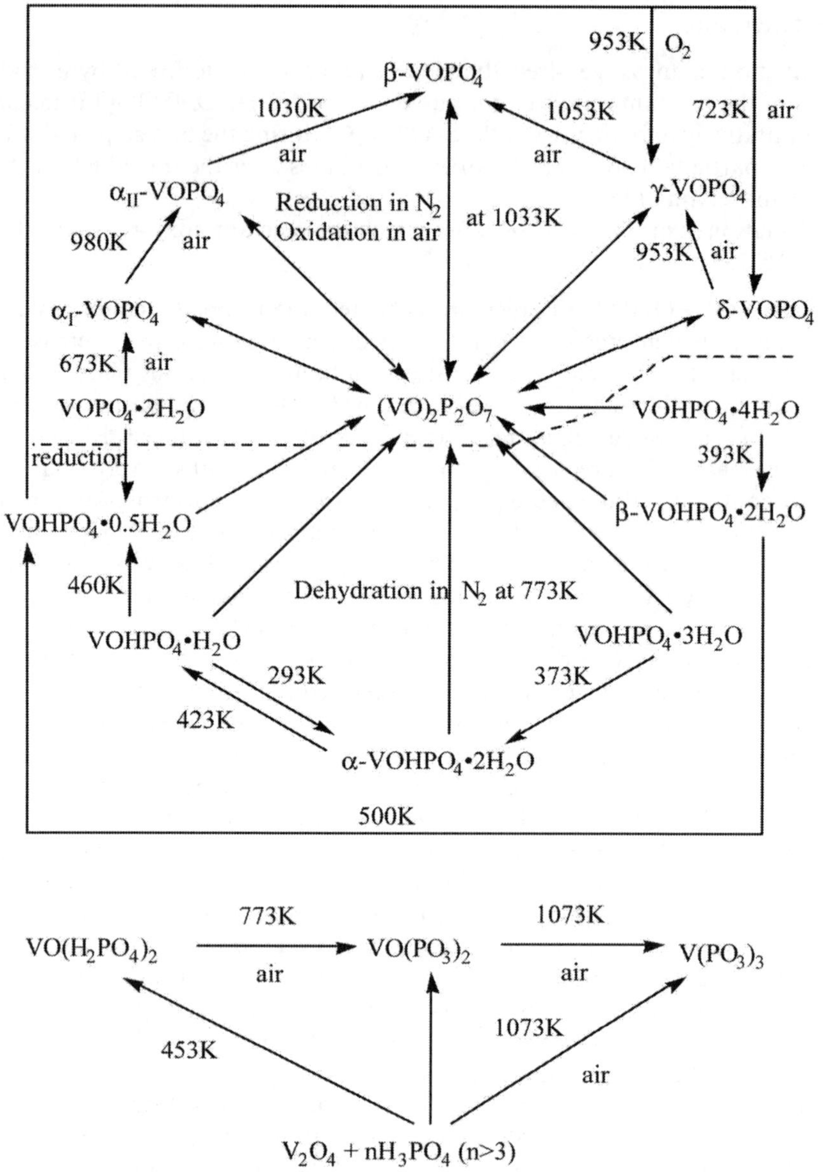

Figure 2 *Phase transformations in the VPO system [43]. Dashed lines separate the reversible redox transformations to* $(VO)_2P_2O_7$ *(top) from irreversible dehydrations (bottom)*

with the preservation of the V-O-P connectivity.

Thompson *et al.* [46] analyzed symmetries of the two structures and suggested that this transformation is unlikely topotactic. Their suggestion is in agreement with the observation of an amorphous intermediate phase during the transform-

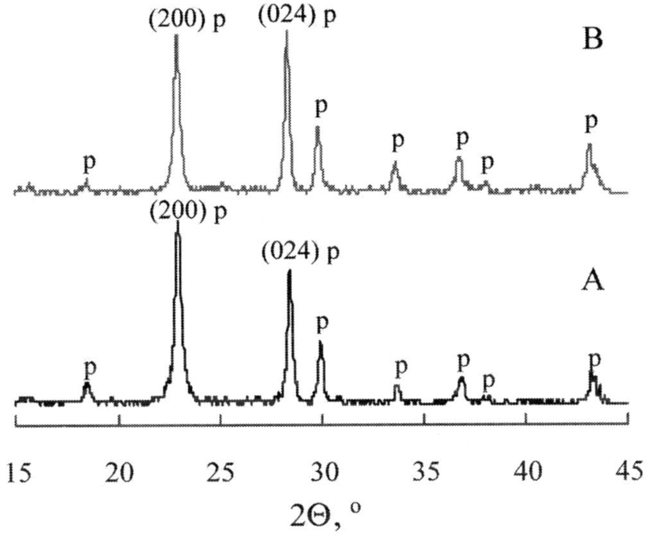

Figure 3 *XRD patterns of equilibrated organic (A) and aqueous (B) VPO catalysts [63]. The same system of labeling as in Figure 1. Peak width: FWHM of (200)= 0.35 and 0.34°; FWHM of (024)= 0.25 and 0.22° in organic and aqueous catalyst, respectively; Relative peak intensity: I(200)/I(024)= 1.41 and 0.93 in organic and aqueous catalyst, respectively*

ation of $VOHPO_4{\cdot}0.5H_2O$ to $(VO)_2P_2O_7$.

Torardi *et al.* [47] studied the thermal transformation of the $VOHPO_4{\cdot}0.5H_2O$ precursor into $(VO)_2P_2O_7$ by electron and X-ray diffraction techniques and showed that the transformation was topotactic in a sense that the initial crystal morphology was preserved during the transformation. Single crystals of $VOHPO_4{\cdot}0.5H_2O$ were converted to pseudomorphs, which were unchanged in size or shape with respect to the starting crystals. To account for these observations, Torardi *et al.* [47] proposed a detailed topotactic phosphorus-inversion mechanism for this transformation. According to this mechanism, the coordination water molecules are first lost to form edge-sharing vanadyl dimers with concomitant decrease in the *d*-spacing. The vanadyl dimers then pivot about the shared edge to a parallel arrangement of V=O bonds. This rotation results in ca. 12% expansion of the *a*-lattice parameter. This is followed by proton transfer from half of the $[HPO_4]$ groups to yield $[PO_4]$ and $[H_2O\text{-}PO_3]$ entities. The subsequent loss of water from the latter units sets stage for full condensation of the layers into a three-dimensional network. The square-pyramidal VO_5 pairs of adjacent layers connect via the displacement of one of the V atoms in a pair along the *c* axis toward an apical oxygen atom in the next layer. This creates edge-sharing vanadyl dimers with *trans* arrangement of vanadyl oxygens in each dimer along the *c* axis. Finally, highly reactive $[PO_3]$ units invert by movement of the P atom in the *c* direction through the plane of the three basal oxygen atoms to bond to a $[PO_4]$ unit located above or below the original layer. Torardi *et al.*

[47] explained the appearance of an amorphous intermediate by significant disruption of the $VOHPO_4 \cdot 0.5H_2O$ structure during the transformation to $(VO)_2P_2O_7$ according to the phosphorus-inversion mechanism. The degree of the long-range order during the transformation depends on the presence of structural defects in the $VOHPO_4 \cdot 0.5H_2O$ structure. The organic precursors usually exhibit crystallographic disorder associated with its retained alcohol content, which translates into a similar disorder in the pyrophosphate phase and is likely to be responsible for the appearance of an amorphous intermediate. On the other hand, highly crystalline aqueous precursor was found to transform into $(VO)_2P_2O_7$ without the amorphous intermediate [47]. Both crystalline $VOHPO_4 \cdot 0.5H_2O$ and $(VO)_2P_2O_7$ were found to exist simultaneously during the transformation of an aqueous precursor.

3.1 The Role of Conditioning Procedure. – After the crystalline $VOHPO_4 \cdot 0.5H_2O$ precursor is transformed into the fresh, poorly crystalline $(VO)_2P_2O_7$ and overoxidized V(V) orthophosphate impurity phases in oxygen-containing atmosphere, changes in its catalytic performance and physicochemical properties occur during on-stream conditioning in the reactant mixture. A 'nonequilibrated' catalyst is termed as 'fresh' in contrast to an 'equilibrated' catalyst which spent considerable time, e.g. 30 days, under catalytic reaction conditions [4,10,48]. A fresh catalyst is more active and less selective in *n*-butane activation, particularly at high conversion, owing to the presence of the V(V) orthophosphate impurities and easier oxidation of V^{IV} to V^{V} at the end of the reactor in a fixed-bed operation as the reactant mixture becomes more oxidizing at lower *n*-butane concentration [4,10,48].

The aim of such conditioning procedure is to reduce V(V) orthophosphate impurity phases in the presence of *n*-butane, crystallize $(VO)_2P_2O_7$ in the VPO catalysts and increase their surface area [10]. Several alternative definitions of the equilibrated state of VPO catalysts and models of an active and selective phase and surface sites have been advanced [4, 10-13]. Some of the reasons for the reported differences in the catalytic performance of VPO catalysts may be attributed to:

- Different phase compositions of the VPO catalysts in these studies,
- Only short-term or incomplete kinetic studies conducted on fresh catalyst,
- Limitations of bulk characterization techniques when applied to study catalytic reactions occurring at the surface.

A number of crystalline $VOPO_4$ phases were detected in the VPO catalysts by different research groups [11, 12] that displayed vastly different catalytic behavior during on-stream conditioning (Figure 2). Poorly crystalline and amorphous $VOPO_4$ phases present in the catalysts are often undetected by XRD. The presence of such disordered phases may be confirmed by other characterization techniques, such as Raman spectroscopy and ^{31}P spin echo NMR. The incomplete phase analysis of VPO catalysts is one of the major reasons responsible for several alternative definitions of the equilibrated state and the nature of the active VPO phase [10-12]. In many cases, only short-term VPO conditioning was conducted and the attainment of the steady state was reported after only several hours on-stream, e.g. [49-51]. Limitations of the bulk characterization techniques in studying the catalytic reaction occurring at the surface have been demonstrated in several studies [23, 52, 53]. For example, equilibrated organic catalysts displayed superior performance in *n*-butane oxidation over aqueous catalysts, while the phase composition and crystalline order of $(VO)_2P_2O_7$ in these catalysts after conditioning were almost identical [42] (Figure 3). No significant changes in the bulk phase composition (XRD) or the surface P/V ratios (XPS) have been observed during VPO conditioning.

Gai and Kourtakis [54] investigated the changes occurring in the bulk (200) planes of a vanadyl pyrophosphate catalyst upon prolonged treatment in *n*-butane, nitrogen, and hydrogen at low pressure in a high-resolution transmission electron microscopy (HRTEM) study. Under such exotic reducing conditions a high concentration of anion vacancies was introduced in the bulk (200) plane. To minimize the strain energy associated with these defects, the $(VO)_2P_2O_7$ crystal glided along the <201> direction by a pure shear mechanism creating fault boundaries running from the surface into the crystal bulk. After reduction for a few days, one line defect per ca. 7 unit cells was observed, corresponding to the defective $(VO)_2P_2O_{6.67}$ phase with a vanadium oxidation state of ca. +3.7 [54]. Gai and Kourtakis [54] proposed that these surface line defects were associated with the active and selective sites in *n*-butane oxidation to maleic anhydride in accordance with the model of Centi *et al.* [10]. However, they did not report the kinetic data for the VPO catalysts of their HRTEM study. Therefore, further studies are necessary in order to confirm whether these bulk defects are present in the VPO catalysts under realistic catalytic reaction conditions.

The surface region of the (200) planes of vanadyl(IV) pyrophosphate catalysts was recently studied by means of high resolution electron microscopy [55]. The flat platelet morphology of $(VO)_2P_2O_7$ was particularly suited for a profile HRTEM study of the surface (200) planes (Figure 4) [56], which contain the vanadyl dimers associated with the active and selective surface sites for *n*-butane oxidation [55]. The evolution of the bulk crystallinity as a function of time on *n*-butane/air stream was monitored by XRD (FWHM of the (200) reflection) and Raman spectroscopy (the P-O-P stretch at 921 cm^{-1}) [55]. The steady-state performance in *n*-butane oxidation to maleic anhydride was attained in this study after ca. 10 days on stream and correlated with disappearance of a thin 15 Å amorphous layer covering the surface (200) planes of $(VO)_2P_2O_7$. This amorphous region contracted to ca. 5 Å in the catalyst after 8 days in the reactor, and the (200) planes extended up to the last surface layer in the equilibrated catalysts after 23 days on stream. Only $(VO)_2P_2O_7$ was detected in these VPO catalysts during the entire time on stream by XRD and Raman spectroscopy. The ordering of the $(VO)_2P_2O_7$ surface coincided with much higher steady state selectivity to maleic anhydride. This HRTEM study [55] provided evidence that the crystallization in the bulk occurred on a longer time scale than the ordering in the surface region of $(VO)_2P_2O_7$ and that the catalytic performance correlated with the surface ordering.

These findings further indicated that although different preparation routes led

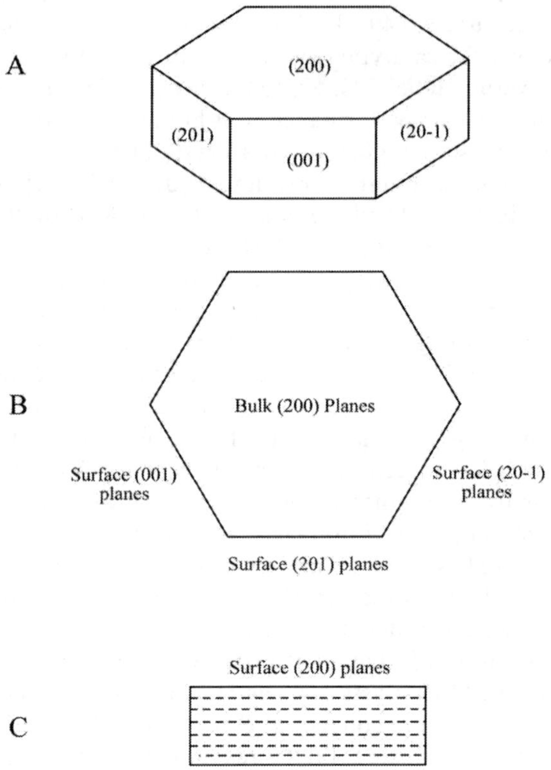

Figure 4 *Exposure of crystal faces in a* $(VO)_2P_2O_7$ *crystallite (a) according to Matsuura and Yamazaki [56] in relation to imaging conditions of HREM experiments of (b) Gai and Kourtakis [54] and (c) Guliants* et al *[55]. The direction of an electron beam in the cases (a) and (b) is normal to the surface of the paper*

to the same bulk vanadyl(IV) pyrophosphate phase, they displayed different relative exposure of the selective (200) planes and different surface termination depending on the synthesis and conditioning methods used [11]. The development of high steady-state selectivity of vanadyl(IV) pyrophosphate catalysts correlated with gradual disappearance of a thin amorphous layer terminating the (200) surface planes.

3.2 Role of the P/V Ratio. – The optimal catalyst composition is characterized by a slight excess of phosphate with respect to the empirical formula of the $VOHPO_4 \cdot 0.5H_2O$ precursor [16, 23, 57]. A considerable excess of surface phosphate (P/V = 1.5-3.0 by XPS [58]) prevents the bulk oxidation of $(V^{IV}O)_2P_2O_7$ to the V^VOPO_4 phases [10, 57, 59]. In catalysts with a slight phosphate deficiency (P/V = 0.95), the rates of V^{IV} oxidation and reduction are high [60]. The increased V^V content leads to more active but less selective catalysts, while low reducibility of V^{IV} in VPO catalysts with high surface phosphate concentration results in low catalytic activity. Therefore, catalysts with a slight excess of phosphate (P/V = 1.05) show the right compromise between reducibility and oxidizability needed to obtain both high activity and selectivity in *n*-butane oxidation. According to Matsuura and Yamazaki [56], the excess phosphate terminates the side faces of the (200) plane of $(VO)_2P_2O_7$ (i.e. 001, 021, 02-1, etc.) in the form of the surface $VO(PO_3)_2$ phase which prevents the oxidation of vanadyl pyrophosphate due to lower oxidizability of $VO(PO_3)_2$. The role of high surface phosphate concentration in VPO catalysis will be discussed in more detail within the context of the active surface.

4 Models of Active VPO Surface

The VPO system is characterized by the existence of various V^{IV}/V^{V} phosphate phases and $(V^{IV}O)_2P_2O_7$ which may readily interconvert depending on the redox properties of the *n*-butane/air feed, the time on stream, and the reaction temperature [10-12]. Although vanadyl pyrophosphate has been identified as the only crystalline phase present in the best VPO catalysts, the V^{V} and V^{IV} phosphate phases can be present in both crystalline and disordered state depending on the particular conditioning procedure used or the degree of catalyst equilibration [10]. This complexity of the solid-state chemistry of the VPO system is perhaps responsible for a number of competing proposals of the active and selective VPO phase reported to date in the catalysis literature.

Bordes [43] suggested that the active sites in *n*-butane oxidation to maleic anhydride are associated with coherent interfaces between slabs of the (100) planes of various $VOPO_4$ phases and the (200) planes of $(VO)_2P_2O_7$ along the (001) and (201) planes, respectively. However, the best $(VO)_2P_2O_7$ catalysts display the lack of other impurity $VOPO_4$ phases. Therefore, the mechanism of Bordes [43] may be appropriate to explain the catalytic behavior of nonequilibrated or overoxidized VPO catalysts that contain various microcrystalline $VOPO_4$ phases.

Volta *et al.* [61], on the contrary, believed that the active sites are not associated with interfaces between these crystalline phases. On the basis of comparison between XRD and radial electron distribution data they suggested that the active phase for selective oxidation of *n*-butane consists of a mixture of well-crystallized $(VO)_2P_2O_7$ and an amorphous $V^{V}OPO_4$ phase involving corner-sharing VO_6 octahedra (Figure 5a). This amorphous phase was interpreted by them as a precursor of *b*-$VOPO_4$, which formed at higher reaction temperatures. More recently, Volta *et al.* [62] suggested on the basis of kinetic data as well as XRD and ^{31}P MAS NMR results that domains of γ-$VOPO_4$ supported on a $(VO)_2P_2O_7$ matrix are necessary for selective *n*-butane oxidation. However, XRD, Raman and ^{31}P NMR studies showed that the best catalysts did not contain amorphous or microcrystalline V^{V} phosphates [63]. Therefore, the mechanism of Volta *et al.* [58,59] may explain the behavior of only nonequilibrated or overoxidized VPO catalysts that contain $VOPO_4$ phases.

Hutchings *et al.* [64] and Volta *et al.* [65] suggested that the actives sites for *n*-butane oxidation to maleic anhydride comprise a V^{4+}/V^{5+} couple well dispersed on the surface of a range of VPO phases, e.g. $(VO)_2P_2O_7$ for equilibrated VPO catalysts. Hutchings *et al.* also included the possibility that the actives sites could be the defects found by Gai and Kourtakis [54] as well as interfaces between microcrystalline $VOPO_4$ phases and well-crystalline $(VO)_2P_2O_7$.

Misono *et al.* [66] observed reversible formation of the surface X1 phase during partial oxidation of $(VO)_2P_2O_7$ and proposed it as an active and selective phase in *n*-butane oxidation. The X1 phase was detected by the appearance of the XRD reflection at 21.6° 2θ and Raman bands at 1090, 1020, and 937 cm^{-1}, the latter band being nearly coincident with the strong P-O-P stretch of $(VO)_2P_2O_7$ [63]. However, the fingerprint XRD and Raman features of the X1 phases closely resemble those of microcrystalline $VOPO_4$ occuring at 21.3° 2θ and 1092 and 1018 cm^{-1}, respectively [63]. Although this model [66] can explain the catalytic behavior of overoxidized VPO catalysts, the presence of the X1 phase is not required for selective oxidation of *n*-butane on equilibrated VPO catalysts that show the presence of $(VO)_2P_2O_7$ only.

Yamazoe *et al.* [49] have reported $VO(H_2PO_4)_2$ as the precursor of the active and selective phase in *n*-butane oxidation. This precursor transformed to an amorphous $VO(PO_3)_2$ catalyst which was much less active but just as selective as the $(VO)_2P_2O_7$ catalysts (Figure 5b). Volta *et al.* [67] have recently found traces of $VO(PO_3)_2$ in the VPO catalysts by ^{31}P NMR spin echo mapping. They concluded that the V^{5+} species in strong interaction with the $VO(PO_3)_2$ structure is a controlling factor in the catalytic performance in *n*-butane oxidation to maleic anhydride. However, low activity and high selectivity of $VO(PO_3)_2$-containing catalysts in *n*-butane oxidation was later explained by the presence of either $(VO)_2P_2O_7$ impurity or $VOPO_4$ phases formed during solid state disproportionation of $VO(PO_3)_2$ into V(III) and V(V) phases [63].

Trifirò *et al.* [10] attributed the activity of the VPO catalysts in *n*-butane oxidation to vanadyl pyrophosphate, whereas the selectivity to maleic anhydride was associated with the presence of a very limited and controlled amount of V^V sites (Figure 5c). TEM analysis of spent catalysts, as well as the ratio of intensities of the interlayer (200) to in-plane (024) reflection in the XRD patterns of $(VO)_2P_2O_7$ catalysts showed a high exposure of the (200) planes in the $(VO)_2P_2O_7$ crystallites. Trifirò *et al.* [10] suggested that the active surface is obtained by truncation of the $(VO)_2P_2O_7$ crystals along the (200) plane. Other studies of well-defined model VPO catalysts showed that the best catalysts contained only well-crystalline $(VO)_2P_2O_7$ [4]. No other microcrystalline or disordered impurity phases were detected in such catalysts [4]. The attainment of the steady-state performance coincided with disappearance of a thin amorphous layer present in the surface region of the (200) planes of $(VO)_2P_2O_7$ [4]. The active and selective surface sites for *n*-butane oxidation to maleic anhydride were associated with the presence of vanadyl dimers in the surface (200) planes of $(VO)_2P_2O_7$ [4].

On the other hand, the side (021) and (001) planes were proposed by Matsuura and Yamazaki [67] and then later by Volta *et al.* [68] and Okuhara *et al.* [69] as the surfaces unselective in *n*-butane oxidation (Figure 6).

Cavani and Trifirò [13] further described four types of termination of the

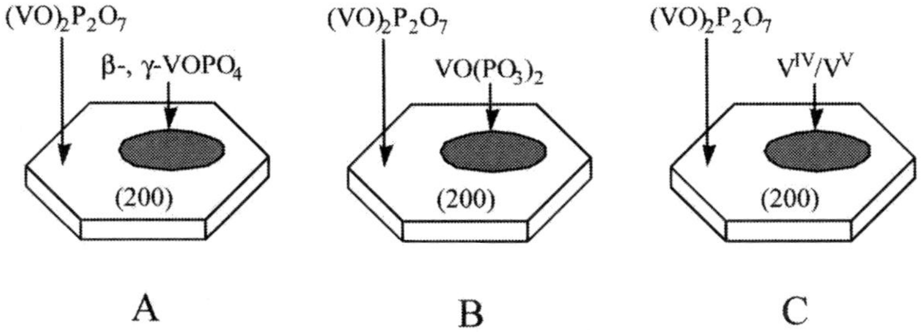

Figure 5 *Models of active surface of the VPO catalysts according to (***A***) Volta* et al. *[61, 62], (***B***) Yamazoe* et al. *[49] and (***C***) Trifirò* et al. *[10].*

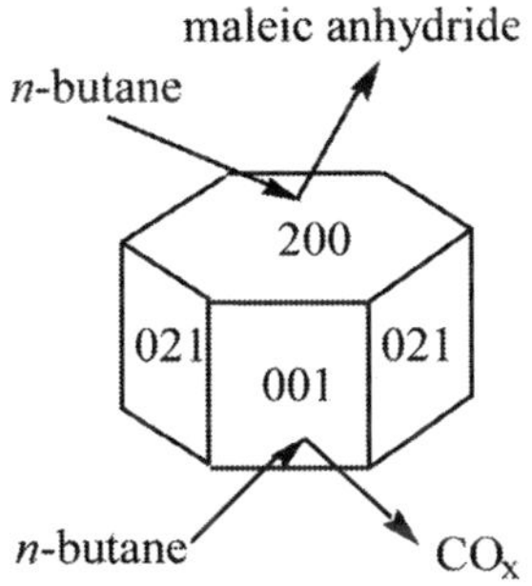

Figure 6 *Selective and non-selective planes of vanadyl pyrophosphate for n-butane oxidation [10, 49, 61, 62]*

surface (200) planes in $(VO)_2P_2O_7$: A, B, C, and D. Type A termination where only vanadium species are present on the surface is the least probable according to Ebner and Thompson [14], since excess phosphate is always present in real catalysts. Such situation would be characterized by high activity and low selectivity to maleic anhydride [58], in agreement with the catalytic behavior of VPO catalysts with a $P/V < 1$.

Ebner and Thompson [58] proposed the termination of the surface with pendant $H_2P_2O_7^{2-}$ groups, which surround two vanadyl dimers (type B). Such surface termination may be envisioned by replacing every phosphate tetrahedron facing up in the bulk (200) planes of $(VO)_2P_2O_7$ with pyrophosphate groups. In this type of surface termination, the pyrophosphate 'fence' isolates pairs of vanadyl dimers from other pairs, which was proposed to be beneficial for the selectivity to maleic anhydride during *n*-butane oxidation.

In the surface (200) planes terminated with HPO_4^- groups (type C), every phosphate tetrahedron facing up in the bulk (200) planes of $(VO)_2P_2O_7$ would correspond to a surface HPO_4^- group. This situation is similar to that of V^VOPO_4, and the vanadium sites in the dimers are more easily accessible from the gas phase than in the type B termination, which is expected to be detrimental for the selectivity to maleic anhydride. In such case the formation of discrete V^V species or V^VOPO_4 phases can occur [69]. The (200) surface termination with

$VO(PO_3)_2$ (type D) occurs when considerable excess of phosphate is present [68], which significantly lowers the catalytic activity in *n*-butane oxidation.This situation corresponds to the above-mentioned model of active surface proposed by Yamazoe *et al.* [49] and shown in Figure 5b.

5 Proposed Steps in *n*-Butane Oxidation to Maleic Anhydride

The reaction steps proposed for the transformation of *n*-butane to maleic anhydride [4] are shown in Table 1. However, none of the intermediates shown in Table 1 was detected under typical reaction conditions for kinetic reasons.

The activation of secondary C-H bonds in *n*-butane is the first and rate-determining step. When compared to oxidation of 1-butene on the VPO catalysts, the rate of *n*-butane disappearance is 20 [66] to 60 [10] times slower. This may result in a rapid transformation of the intermediates to maleic anhydride before any desorption of the former may occur.

The following experimental observations led to the proposed reaction steps shown in Table 1:

- 1-Butene, *cis*/*trans*-2-butenes, butadiene and furan have been detected in the oxidation of *n*-butane on the VPO catalysts under very unusual conditions, such as under oxygen deficiency at high *n*-butane concentration and very short contact times [9], or in high vacuum in a temporal analysis of products (TAP) reactor [70].
- The intermediates in Table 1 (with the exception of 2,5-dihydrofuran and γ-butyrolactone) have been observed in the oxidation of butenes and butadiene to maleic anhydride on the VPO catalysts [10].
- 2,5-dihydrofuran oxidation to furan occurs over VPO catalysts at much lower temperatures (533K) with very high selectivity (ca. 90 mol. %), which is probably why 2,5-dihydrofuran has not been observed experimentally during C4 hydrocarbon oxidation over the VPO catalysts [10].

The unique features of the VPO catalysts in carrying out the reaction steps shown in Table 1 are (i) the ability to selectively activate *n*-butane during the rate-determining step, (ii) rapid oxidation of chemisorbed intermediates to maleic anhydride with high selectivity, and (iii) the lack of desorption of any intermediate contributing to high selectivity to maleic anhydride. Other cata-

Table 1 *Proposed reaction steps in n-butane oxidation to maleic anhydride*

Step	*Reaction*
n-butane ⇒ butenes	oxidative dehydrogenation
butenes ⇒ butadiene	allylic oxidation
butadiene ⇒ 2,5-dihydrofuran	1,4-oxygen insertion
2,5-dihydrofuran ⇒ furan	allylic oxidation
furan ⇒γ-butyrolactone	electrophilic insertion
γ-butyrolactone ⇒ maleic anhydride	electrophilic oxygen insertion

lysts, such as the MoO_3-based mixed oxides [10, 70], are active but unselective in *n*-butane oxidation. Due to the high rate of oxidation of intermediates on VPO catalysts, their surface concentration is low which makes it difficult to detect these intermediates by *in situ* IR spectroscopy [10].

The reaction pathway (Table 1) shows the single steps and the polyfunctional nature of active sites that are required for the oxidation of *n*-butane to maleic anhydride [71-74]. The polyfunctional nature of the VPO catalysts has been further demonstrated by the oxidation of probe molecules (Table 2) in accordance with the steps proposed for *n*-butane oxidation to maleic anhydride [13, 75, 76, 77]. The high yield of maleic anhydride observed in the oxidation of these molecules provided support for the reaction steps shown in Table 1.

The proposed reaction pathway shown in Table 1 has been recently challenged by Chen and Munson [78,79]. They employed a series of fresh VPO catalysts including one $(VO)_2P_2O_7$ and a number of overoxidized amorphous VPO phases possessing the bulk P/V ratios of 0.9-1.2, the average V oxidation states in the 3.92-4.95 range and ^{13}C-labeled *n*-butane reactant in the absence of dioxygen under unconventional pseudo-flow reaction conditions. They observed that the label in the butadiene, produced from 1,4-^{13}C-*n*-butane, was completely scrambled, but in maleic acid, also produced from 1,4-^{13}C-*n*-butane, it was largely unscrambled. They concluded that butadiene was unlikely to be the primary reaction intermediate for the conversion of *n*-butane to maleic anhydride, because of the observed discrepancy in the amount of label scrambling between maleic acid and butadiene. However, the small amount of label scrambling in maleic acid suggested that both reaction pathways may occur simultaneously to produce maleic acid. Chen and Munson further observed formation of much more isotopically labeled ethylene from fully ^{13}C-labeled butane as from [1,4-^{13}C]butane, indicating that ethylene was produced mainly from the two methylene carbons of *n*-butane, while the abstraction and oxidation of the two methyl groups of *n*-butane resulted in the total oxidation of *n*-butane. However, further studies employing well-characterized equilibrated VPO catalysts under realistic reaction conditions are necessary to investigate an organometallic mechanism of *n*-butane oxidation to maleic anhydride proposed by Chen and Munson [78,79].

Table 2 *Polyfunctional nature of the VPO catalysts.*

Reaction	*Probe molecule*	*Product*
oxidative dehydrogenation	isobutyric acid	methacrylic acid
	cyclohexane	benzene
allylic oxidation	butene	butadiene
	2,5-dihydrofuran	Furan
1,4-oxygen insertion	benzene	maleic anhydride
	naphthalene	naphthaquinone
electrophilic oxygen insertion	methacrolein	methacrylic acid
	o-xylene	phthalic anhydride

6 Proposed Active Sites and Mechanisms of *n*-Butane Oxidation

The best VPO catalysts preferentially expose the (200) planes of $(VO)_2P_2O_7$ [4, 10-13]. Therefore, the models of *n*-butane oxidation on the VPO catalysts proposed to date are based on the hypothetical active sites present on the surface (200) plane. The following types of hypothetical active sites have been proposed to exist on the surface (200) plane [13]: (i) Brønsted acid sites, probably -POH groups; (ii) Lewis acid sites, probably V^{IV} and V^{V}; (iii) one electron redox couple, V^{V}/V^{IV}, V^{IV}/V^{III}; (iv) two electron redox couple, V^{V}/V^{III}; (v) bridging oxygen, V-O-V, V-O-P, or VO(P)V; (vi) terminal oxygen, $V^{V}{=}O$, $V^{IV}{=}O$; and (vii) activated molecular oxygen, θ^1-peroxo and θ^2-superoxo species. The roles of these species according to the proposed mechanisms of *n*-butane oxidation are discussed below.

Pepera *et al.* [80] studied the oxidation of *n*-butane to maleic anhydride on a $(VO)_2P_2O_7$ catalyst using labeled compounds, such as D_2O, $^{18}O_2$, 2,2,3,3-D_4-*n*-butane, and 1,1,1,4,4,4-D_6-*n*-butane. They found that every two surface vanadium atoms are capable of activating one molecule of oxygen, while the bulk of the catalyst did not participate in *n*-butane oxidation. *n*-Butane does not reversibly chemisorb with hydrogen exchange onto the surface, while maleic anhydride does. The surface of the oxygen-equilibrated catalyst does not undergo exchange with the gas phase oxygen, indicating the irreversibility of the surface reoxidation step. The primary kinetic isotope effects observed in deuterated *n*-butane studies strongly suggested that the oxidation of *n*-butane both to maleic anhydride (activation of methylene C-H bonds, 95% of all *n*-butane activated) and to combustion products (methyl C-H bond activation, 5% of all *n*-butane activated) proceeds through the same rate-determining step. Selectivity to maleic anhydride is determined during the fast steps after the initial C-H bond activation. The kinetic isotope effect observed ($k_H/k_D = 2.18$) is close to $k_H/k_D = 2.27$ expected for the complete breaking of the C-H(D) bond in the transition state with little or no concomitant creation of another bond at 673K. The latter value is based on the $\delta\Delta G^{\ddagger} = 365\ cm^{-1}$ (1.05 kcal) for the cleavage of a C-H vs. a C-D bond of an sp^3-hybridized carbon [68]. Recently, Wang and Barteau [81-83] studied the kinetics of *n*-butane oxidation with a novel oscillating microbalance reactor. They found that the reduction rate of equilibrated VPO catalyst with butane was 0.4th order in the *n*-butane concentration and 4th order in the available lattice oxygen concentration. This proposed rate law represented the reduction of the catalyst to an extent of the equivalent to the removal of several layers of lattice oxygen. According to these results, four surface oxygen ions could be involved in the rate-determining step for *n*-butane oxidation (i.e. activation of *n*-butane). Furthermore, they found [81-83] that the oxidation of *n*-butane by lattice oxygen accounted for only 5 % of the *n*-butane oxidation rate, while the rest was carried out by adsorbed oxygen (most likely O^-). The determination of surface and lattice oxygen mobility in VPO catalysts has been also studied by Volta *et al.* [84].

Pepera *et al.* [80] further argued that an active surface V^{IV} site could be viewed as a transition-metal radical. While highly active oxygen species [85] would show selectivity for the C-H bonds in *n*-butane only in a statistical sense [86] (i.e. only 40% selectivity to maleic anhydride), V^{IV} due to lower reactivity of its $d_{x^2-y^2}$ orbital based unpaired electron would discriminate between the methylene and stronger methyl C-H bonds. Although Pepera *et al.* [80] did not present the mechanism of oxidation, they concluded that V^{IV} on the surface of $(VO)_2P_2O_7$ may assist homolytic cleavage of the methylene C-H bond of *n*-butane:

$$V^{IV} + RH \longrightarrow R\bullet + V^{III}$$

Centi *et al.* [10] have compared the rate constants for depletion of th·e C_2-C_7 alkane series on a $(VO)_2P_2O_7$ catalyst with the kinetic constants [87] for the theoretical reaction of simultaneous abstraction of two hydrogen atoms and obtained a linear correlation. Their results supported a hypothesis that the rate-determining step is the contemporaneous removal of two methylene hydrogen atoms from the carbon atoms in the 2- and 3-positions in *n*-butane. Centi *et al.* [10] proposed that the Lewis acid site and the bridging oxygen abstract two hydrogen atoms from the two methylene groups of *n*-butane via a concerted mechanism shown in Figure 7.

The hydrogen atom abstraction by the V^{4+} Lewis acid site in the model of Centi *et al.* [10] may be broken up into the following steps:

$$C_4H_{10} = C_4H_9\bullet + H\bullet, \qquad \Delta H^\circ = +\ 411\ \text{kJ/mol}\ [88]$$
$$V^{4+} + H\bullet = (VH)^{4+}, \qquad \Delta H^\circ = -200\ \text{kJ/mol}\ [88]$$

The heats of the C-H bond dissociation in alkanes with charge separation [85] in the gas phase are considerably higher than for the homolytic process (ca. 1600 kJ/mol vs. 411 kJ/mol). The enthalpy change associated with the V-H bond formation may be estimated in a following way. Beauchamp and Armentrout *et al.* [89-91] have studied activation of alkanes by gas phase metal and oxometal ions using ion beam reactive scattering technique and determined the energies for the gas phase heterolytic dissociation of metal-hydrogen, metal-hydroxyl and metal-carbon bonds. For most transition metal ions studied, including V-H, the

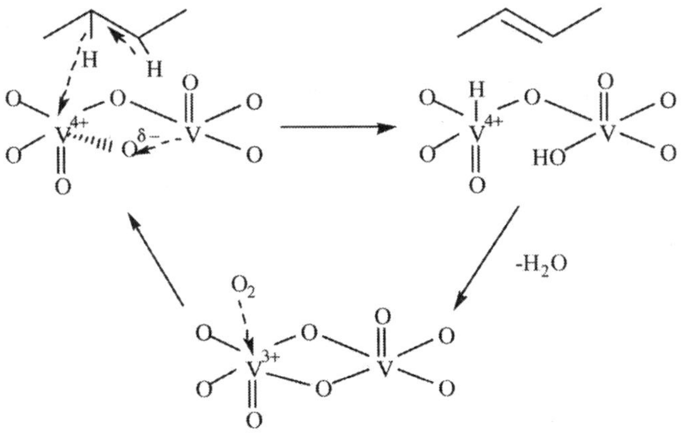

Figure 7 *Mechanism of* n-*butane activation on* $(VO)_2P_2O_7$ *proposed by Centi* et al. *[10]*

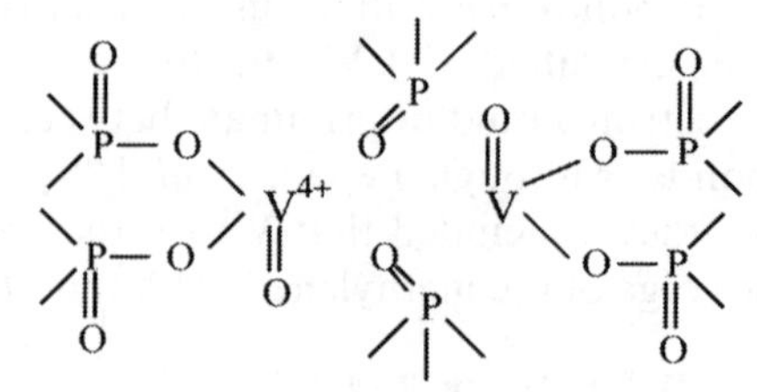

Figure 8 *Hypothetical VPO connectivity in the surface (200) plane in accordance with the elemental composition and charge balance in* $(VO)_2P_2O_7$ *[103]*

homolytic dissociation energy D_o did not exceed 200 kJ/mol. Assuming that the ΔH^o for the V-H bond formation in the model of Centi *et al.* [10] has an upper limit of -200 kJ/mol, the overall process of the hydrogen atom abstraction by V^{4+} is thermodynamically unfavorable ($\Delta H^o = +211$ kJ/mol).

The structure of the surface (200) planes terminated according to the above-mentioned types A-C would be remarkably different from that of the bulk (200) planes of $(VO)_2P_2O_7$. According to the conventional organometallic approach, integer formal charges and oxidation states are assigned to reactants to follow the oxidation state changes in order to understand the nature of elementary reaction steps involved [92]. Therefore, in the model (200) plane of $(VO)_2P_2O_7$ the two-coordinate oxygen corresponds to a single P-O-V bond, while three-coordinate oxygen in P-O-V_2 does not form a formal bond to vanadium atom and corresponds to -P=O in the pyrophosphate anion as shown in Figure 8.

On the other hand, the model of Centi *et al.* [10] shows that the three-coordinate oxygen is involved in the hydrogen atom abstraction leading to the formation of a single -V-OH bond. According to our representation, this situation would correspond to the reaction below:

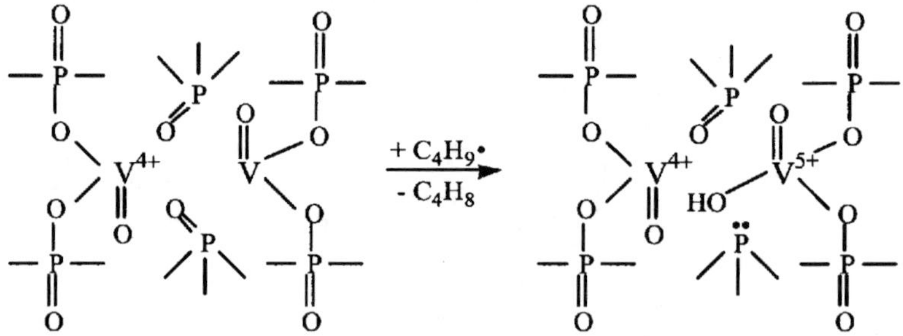

which may be again divided into the following steps:

$C_4H_9\bullet = C_4H_8 + H\bullet$, $\Delta H^o = +411$ kJ/mol [88, 93]
$P^{5+}{=}O = P^{3+} + \bullet O\bullet$ $\Delta H^o = +544$ kJ/mol [94]
$\bullet O\bullet + H\bullet = HO\bullet$ $\Delta H^o = +40$ kJ/mol [91]
$V^{4+} + \bullet OH = V^{5+}OH$, $\Delta H^o = -305$ kJ/mol [95]

ΔH° for the last step may be estimated from the homolytic M-OHdissociation energies for transition metal ions in gas phase determined by ion scattering experiments [86]. The total ΔH° for this process is +690 kJ/mol, and the overall ΔH° associated with the abstraction of two methylene hydrogen atoms (Figure 5) according to model of Centi *et al.* [10] is extremely endothermic (ca. +900 kJ/mol), as was also suggested by Kung [85]. Although Centi *et al.* [10] did not provide a complete mechanism of *n*-butane oxidation to maleic anhydride, they indicated that the Brønsted acid sites may be involved in the steps following the initial activation of *n*-butane.

The Brønsted acid sites were detected by IR spectroscopy and attributed to the presence of P-OH groups belonging to terminal HPO_4^- and $H_2P_2O_7^{2-}$ species [58]. The P-OH groups may have the following functions: (i) to facilitate the removal of water formed during the partial oxidation, (ii) to stabilize the reaction intermediates by forming the surface phosphate esters (P-O-C bonds) and avoiding desorption of these intermediates [96], and (iii) to facilitate the desorption of maleic anhydride preventing its overoxidation [97].

Schiøtt and Jørgensen [98] adopted the frontier orbital approach with extended Hückel calculations to describe formation of 2,5-dihydrofuran from the butadiene intermediate and its oxidation to maleic anhydride on a vanadyl dimer present in the (200) plane of $(VO)_2P_2O_7$ (Figure 9). Their calculations showed that V^{4+}=O is involved in the [2+4]-like cycloaddition of butadiene, which then rearranges to 2,5-dihydrofuran. Molecular oxygen adsorbed on the adjacent vanadium atom in the dimer (the η^1-superoxo or η^2-peroxo species) then activates the C-H bond in the 2-position of 2,5-dihydrofuran, leading to a hydrogen atom transfer to the peroxospecies to give a surface-bound hy-

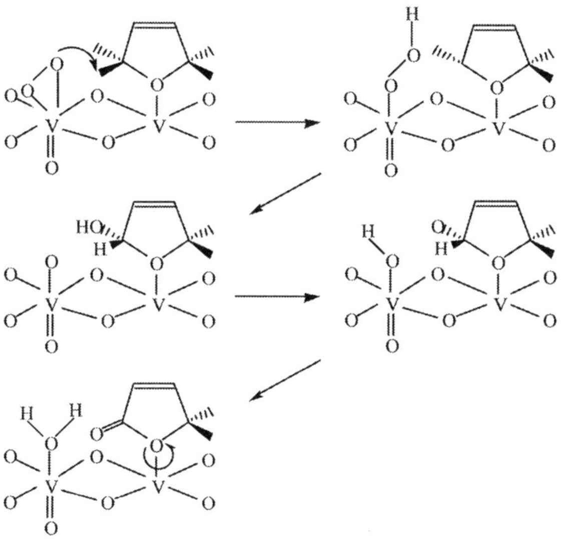

Figure 9 *Schematic representation of oxidation of n-butane on the (200) plane of* $(VO)_2P_2O_7$ *according to Schiøtt and Jørgensen [98]*

droperoxide group. The OH group in -O-O-H then transfers to the neighboring 2,5-dihydrofuran derivative yielding the 2-hydroxy derivative. The asymmetric lactone (γ-butyrolactone) may be obtained by the hydrogen atom transfer to the adjacent radical oxovanadium site. The oxidation of the 5-position of γ-butyrolactone to maleic anhydride may take place in a similar fashion following the desorption of water and activation of another molecule of oxygen on the adjacent reduced vanadium site.

The model of Schiøtt and Jørgensen [98] shows that the proximity of vanadium atoms in a dimer present on the (200) plane of $(VO)_2P_2O_7$ and η^1-superoxo or η^2-peroxo species chemisorbed on coordinatively unsaturated vanadium site are required for the selective oxidation of hydrocarbons. In fact, such activated molecular oxygen species are responsible for the activity and selectivity of vanadium catalysts in homogeneous oxidations [99-101]. Such oxidations are one-electron processes and involve free radicals. Although Schiøtt and Jørgensen [98] did not provide the complete mechanism of *n*-butane oxidation to maleic anhydride on vanadyl dimers, their model indicated that the superoxo species and one-electron, free radical processes may be involved in such oxidation.

Grasselli *et al.* [75] proposed a mechanism of *n*-butane transformation to 1,3-butadiene at the active sites present on the (200) plane of vanadyl pyrophosphate based on conclusions of Schiøtt and Jørgensen [98]. Grasselli *et al.* argued that the dimeric active sites form clusters on the surface, each composed of four vanadyl dimers which are isolated from other clusters by terminal pyrophosphate groups in type B of the surface termination. Chemisorbed oxygen can hop from one site to another within such cluster transforming the state of the active sites. These pyrophosphate groups $[O_3P\text{-}OPO_3H_2]^{2-}$ serve as diffusion barriers,

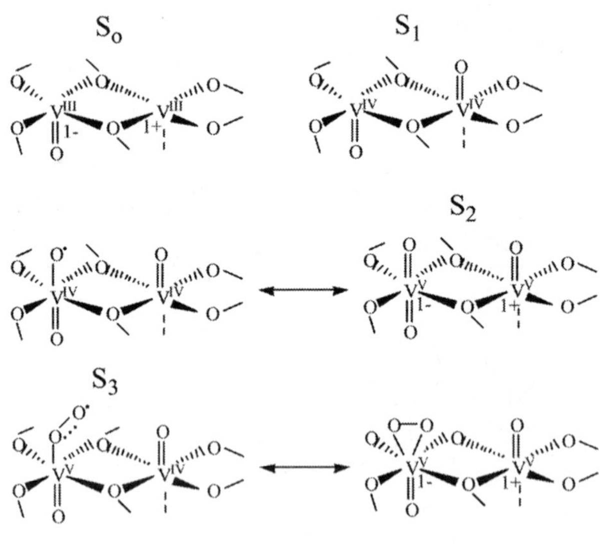

Figure 10 *Schematic representation of the dynamic states S_o-S_3 of the vanadyl dimer present in the (200) plane of $(VO)_2P_2O_7$ according to Grasselli* et al. *[75]*

preventing overoxidation of the reactive surface-bound intermediates by the excess oxygen from neighboring clusters. Such 'site isolation' is considered to be crucial for selective oxidation of hydrocarbons on heterogeneous catalysts [102]. The $[O_3P\text{-}OPO_3H_2]^{2-}$ groups are also Brønsted acid sites and participate in the overall mechanism of oxidation.

Grasselli *et al.* [75] based their model on four vanadium dimers within each cluster that can assume one of the four distinct states, S_o•S_3, shown Figure 10. The original state of the vanadyl dimer (S_1) is that in the bulk (200) plane of vanadyl pyrophosphate. The S_1 site is transformed into S_3 after molecular oxygen forms peroxo or superoxo species on the coordinatively unsaturated V^{IV} in the dimer. Two additional states are possible, denoted by S_o and S_2, with zero and two vanadyl oxygens, respectively.

Grasselli *et al.* [75] proposed that a superoxo species in S_3 activates *n*-butane by abstraction of one of the methylene hydrogen atoms. This results in a surface-bound hydroperoxy group with a simultaneous capture of the alkyl radical by the adjacent vanadyl group to give a surface-bound alkoxy group (Figure 11). The hydroperoxy group can then abstract another methylene hydrogen atom in either α or β-position, resulting in a metal-bound S_1-ketaloxy or S_2-glycoloxy group, respectively (Figure 11).

According, to Grasselli *et al.* [75] the next step is an acid catalyzed dehydration reaction with participation of $[O_3P\text{-}O\text{-}PO_3H_2]^{2-}$ groups, leading to formation of 1,3-butadiene and regeneration of the S_1 site (Figure 12).

The dimeric site in the S_1 state can then activate another dioxygen and oxidize 1,3-butadiene to maleic anhydride in accordance with the model of Schiøtt and Jørgensen [98]. Maleic anhydride desorbing from the surface leaves the active site in the S_o state. In this state the dimer can react with dioxygen to form the S_2 state or accept an oxygen atom from an adjacent site in the S_2 state to yield two dimers in the S_1 state. Grasselli *et al.* [75] proposed a scheme for *n*-butane oxidation to maleic anhydride in the absence of gas phase oxygen involving the four active sites:

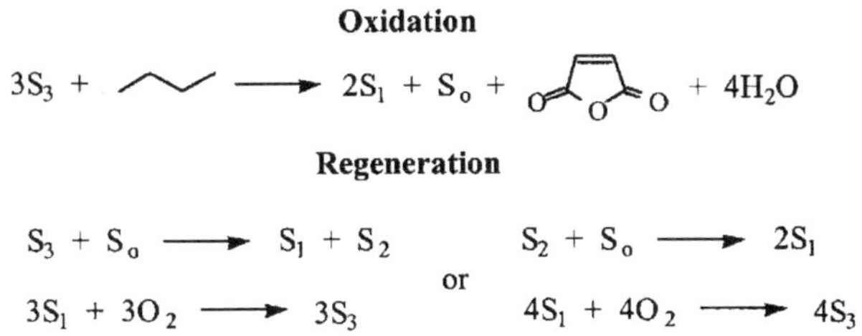

Although the model of Grasselli *et al.* [75] showed the role of superoxo species in *n*-butane activation similar to the model of Schiøtt and Jørgensen [98] and Brønsted acid sites in the 1,3-butadiene formation, it leads to unconventional oxidation states of vanadium as in the case of the model of Centi *et al.* [10]. Grasselli *et al.* [75] (Figure 10) indicated that in the S_3 dimer both vanadyl and peroxovanadyl species are in the +5 state after one electron transfer According to a representation of a vanadyl dimer at the surface (200) plane (Figure 8), the

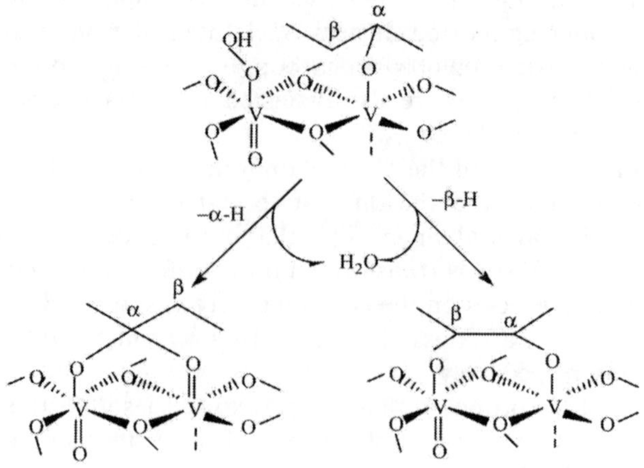

Figure 11 *Activation of n-butane on a vanadyl dimer site in state S_3 according to Grasselli* et al. *[75]*

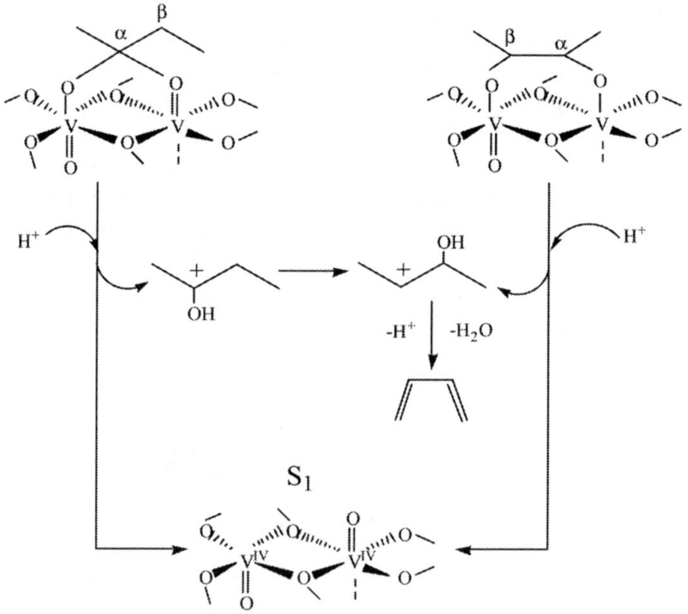

Figure 12 *Acid catalyzed conversion of surface-bound ketaloxy and glycoloxy intermediates to 1,3-butadiene according to Grasselli* et al. *[75]*

peroxo species in the S_3 state in Figure 10 should be coordinated to a vanadium atom in +6 state as it has four single bonds and one double bond. Similarly, after one electron transfer in the case of S_2, Grasselli *et al.* [75] show the formation of $O{=}V^{V}{=}O$ dioxospecies. However, that vanadium atom forms four bonds, two single and two double, and is, therefore, in +6 oxidation state. Vanadyl in S_o according to Grasselli *et al.* [75] is in +3 state despite the fact that it forms two single bonds and one double bond and therefore should be in +4 state.

Guliants and Holmes [103] reported the results of oxidation of various C_4 probe molecules on the equilibrated organic VPO catalyst which showed the presence of only $(VO)_2P_2O_7$ by XRD and Raman. The oxidation of linear and branched hydrocarbons, alcohols, diols and other functionalized C_4 molecules demonstrated the polyfunctional nature of the VPO catalysts and their ability to selectively oxidize various functions to maleic anhydride. They explained formation of various oxidation products on the basis of the structural features of the (200) plane of $(VO)_2P_2O_7$ and the redox chemistry of V(IV) and V(V) species. Guliants *et al.* [104,105] found that the *n*-butane oxidation and maleic anhydride formation rates were enhanced when multiple adjacent surface vanadia sites in model supported vanadia catalysts are present (Figure 13). This observation that the multiple sites are more efficient in *n*-butane oxidation is in agreement with the hypothetical models of Centi *et al.* [10], Grasselli *et al.* [75] and Schiøtt and Jørgensen [98] involving vanadyl dimers in the VPO lattice. The proposed models of *n*-butane oxidation on the VPO catalysts [10, 75, 98] are based on the dimers of edge-sharing *trans*-oxovanadium(IV) octahedra present in the bulk (200) plane of $(VO)_2P_2O_7$. According to the models of Schiøtt and Jørgensen [98] and Grasselli *et al.* [75], the electrophilic vanadyl(IV) oxygen participates in the initial methylene C-H bond activation in *n*-butane and it is subsequently lost during the proposed cycloaddition of the butadiene intermediate leading to the formation of 2,5-dihydrofuran [4] . However, the review of the redox chemistry of vanadium in higher oxidation states [4] revealed only a few examples of such processes involving the vanadyl oxygen. Conversely, the generation of highly reactive vanadium(IV) and (V) η^1-superoxo or η^2-peroxo species in the presence of molecular oxygen is well known [1-4, 46] and widely used in both stoichiometric and homogeneous catalytic oxidations of organic molecules [66] .

Furthermore, the catalytic activity in *n*-butane oxidation over the VPO catalysts was lost when nitrous oxide, a well-known oxotransfer agent, was used instead of molecular oxygen [75]. This suggests that the η^1-superoxo or η^2-peroxo species is indeed involved in the oxidation. In the case of the *trans*-oxovanadium(IV) dimer site proposed as active, the activation of dioxygen

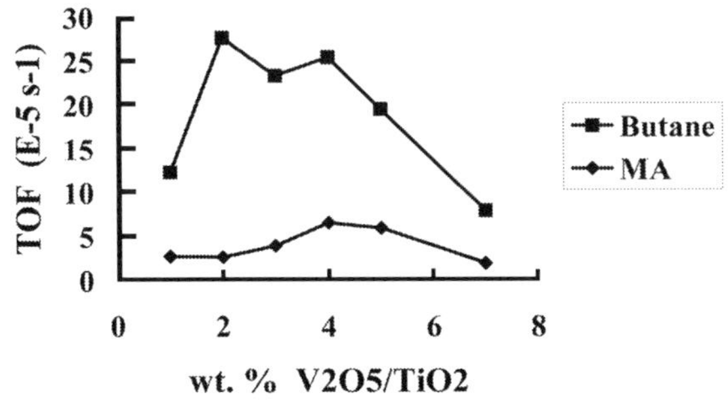

Figure 13 *Oxidation of* n-*butane on supported vanadia catalysts as a function of vanadia surface coverage [105]. Reaction conditions: 1.2 % n-butane in air on V_2O_5/TiO_2 at 494 K*

would lead to only single peroxo species (Figure 10). Meanwhile, the kinetic studies of *n*-butane oxidation suggest a different site capable of contemporaneous abstraction of two methylene hydrogen atoms in *n*-butane [1-4]. In fact, the (200) plane of $(VO)_2P_2O_7$ displays parallel chains of *trans*-oxovanadium(IV) dimers. The *trans*-oxovanadium(IV) dimers are arranged in such a way that the vanadium(IV) atoms with the vacant apical coordination site are always next to each other in a chain. Given the average O-O bond length in vanadium(V) peroxo complexes of 1.5 Å [6], the distance between the oxygen atoms in adjacent peroxo species in a *trans*-oxovanadium(IV) dimer chain (ca. 3.5 Å [30]) would compare favorably with the methylene H-H distance in positions 2 and 3 in the eclipsed conformation of *n*-butane (2.3 Å [43]). These observations led Guliants and Holmes [103] to propose that the *cis*-dimeric site shown in Figure 14, rather than the *trans*-oxovanadium(IV) dimer, may be involved in the C-H bond cleavage in *n*-butane. They further proposed a mechanism of selective oxidation of *n*-butane and other C_4 molecules to maleic anhydride based on such dimeric peroxo site and V^{IV}-V^{V} redox couple [103]. Some of the important observations during oxidation of the probe molecules and the proposed mechanism [103] are described in more detail below on the basis of a such dimeric site.

The oxidation of a branched C_4 alkane, isobutane, was carried out to probe the mechanism of the C-H bond activation of alkanes on the VPO catalysts [103]. Maleic anhydride was among the products of oxidation of this branched alkane. In the case of isobutane **29** (Figure 15), the surface-bound peroxo radical would show discrimination in activating first the weaker tertiary C-H bond. The

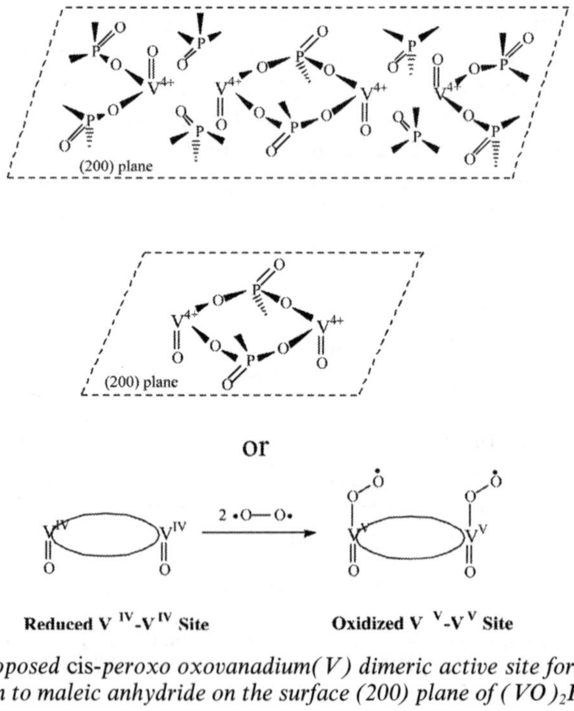

Figure 14 *Proposed* cis-*peroxo oxovanadium(V) dimeric active site for n-butane oxidation to maleic anhydride on the surface (200) plane of* $(VO)_2P_2O_7$ *[103]*

hydroxylation of the *t*-butyl radical would lead to *t*-butanol [4]. Observed formation of methacrolein and methacrylic acid was explained by the allylic oxidation of isobutene obtained by the dehydration of *t*-butanol. The formation of maleic anhydride was accounted for by the skeletal rearrangement of a carbocation or radical intermediate [71]. The *t*-butyl carbocation would be generated in dehydration of *t*-butanol. However, the three methyl groups inductively delocalize the positive charge on the tertiary carbon atom leading to a stable *t*-butyl carbocation. Therefore, the formation of maleic anhydride may not be explained by the rearrangement of *t*-butyl carbocation to linear 2-butyl carbocation. On the other hand, the activation of both the tertiary and the stronger methyl C-H bonds in isobutane by the surface-bound peroxo radical would lead via 1,2-diradical to isobutene **30** and its oxidation products. No products of skeletal rearrangement would be expected in this case either.

The formation of maleic anhydride in the isobutane oxidation was explained by a reaction pathway involving the abstraction of the two hydrogen atoms in positions 1 and 3. Activation of the two methyl C-H bonds may occur resulting in the formation of 1,3-diradical **31**. Simple radicals undergo skeletal rearrangement only when migrating group is aromatic capable of delocalizing an extra electron in the *p*-system. However, in the case of *t*-butyl 1,3-diradical skeletal rearrangement via methyl group migration and a cyclic intermediate (Figure 15) is well documented [71] leading to linear 1,2-diradical **32** and 1-butene **33**. Formation of maleic anhydride during isobutane oxidation on the VPO catalyst suggests that such skeletal rearrangement does occur. On the other hand, it also indicates that the activation of *n*-butane on the VPO catalysts may proceed via

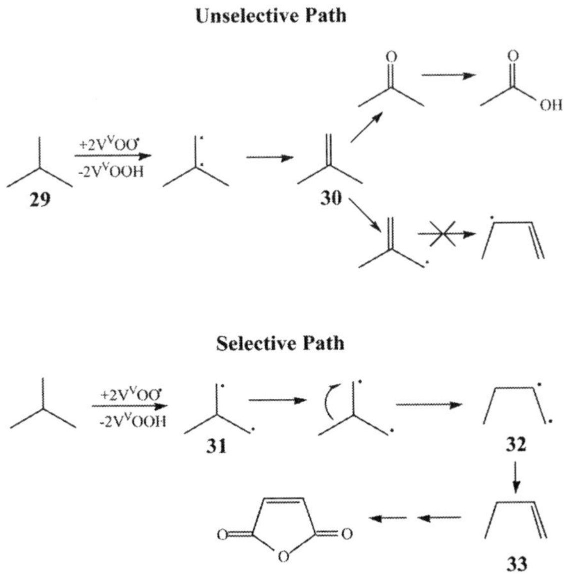

Figure 15 *Proposed oxidation pathway for i-butane on the surface (200) plane of $(VO)_2P_2O_7$ [103]*

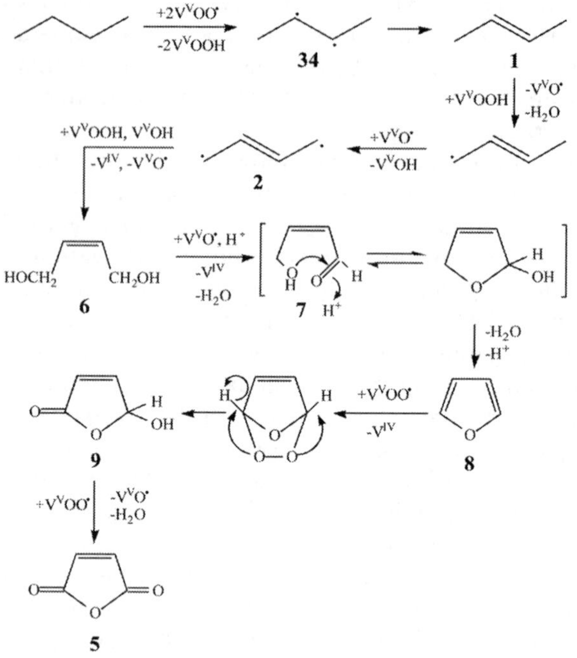

Figure 16 *Proposed mechanism of n-butane oxidation to maleic anhydride over* cis-*peroxo oxovanadium(V) dimers present in (200) planes of $(VO)_2P_2O_7$ [103]*

contemporaneous homolytic C-H bond cleavage and formation of a diradical intermediate.

The proposed mechanism of *n*-butane oxidation on the VPO catalysts (Figure 16) is based on the formation of the 2,3-diradical **34** and further steps of its oxidation to maleic anhydride consistent with the results of oxidation of various C_4 species. 2-Butene undergoes dihydroxylation by abstraction-recombination mechanism via 1,4-alkenyl diradical **2** (Figure 16). Olefinic and *p*-allyl intermediates have been previously detected in the *in situ* IR studies of C_4 hydrocarbon oxidation, including that of *n*-butane, on the VPO catalysts [4]. The unsaturated but-2-ene-1,4-diol **6** thus formed then undergoes oxidation by another V^V-O^- site to yield 4-hydroxy-but-2-enal **7** and V^{IV}. 4-Hydroxy-but-2-enal is dehydrated by the surface HPO_4^{2-} or $H_2P_2O_7^{2-}$ groups to form furan **8** (Figure 16). Furan undergoes cycloaddition with singlet oxygen to yield 5-hydroxy-2(5*H*)-furanones **9** which is oxidized to maleic anhydride by the V^V-O^- species as shown in Figure 15.

7 Effect of Promoters on *n*-Butane Oxidation over VPO Catalysts

The VPO catalysts employed industrially are usually promoted with metal compounds to improve their catalytic performance [4]. The nature, the location and the roles of metal promoters in the VPO system have been previously reviewed [106,107]. However, there is a lack of agreement in the literature concerning the functions of promoters in controlling the morphology, phase and elemental (bulk and surface) composition, structures and redox properties of VPO catalysts, which define their catalytic performance. Moreover, the catalytic performance of both promoted and unpromoted catalysts is frequently reported at low *n*-butane conversion (e.g. 15-50%), while the performance at high *n*-butane conversion (̃90%) is of particular interest for practical applications. Several methods have been used in the past to incorporate promoters in the VPO lattice [106-133]. These methods include (i) the incorporation of promoters during the $VOHPO_4 \cdot 0.5H_2O$ precursor synthesis, (ii) the impregnation of the $VOHPO_4 \cdot 0.5H_2O$ precursor with promoters prior to its thermal activation and (iii) promoter impregnation into VPO catalysts. The promoters can be introduced in the VPO host as metals, salts, oxides and phosphates. Metal phosphates incorporated as ortho- and pyrophosphates have a particularly beneficial effect on catalytic performance [119]. The promoters are typically introduced into VPO catalysts at the 0.25-5 wt. % level [106]. Since the typical surface areas of organic VPO catalysts are in the 20-30 m^2/g range, these levels of promoters correspond to the surface coverage of ~0.1 to 2 theoretical monolayers [105]. At loadings below a theoretical monolayer (~2-3 wt. %), a promoting element may form a two-dimensional surface oxide overlayer and (partially) dissolve in the VPO lattice. The latter is typically observed when soluble promoter salts are introduced directly during the $VOHPO_4 \cdot 0.5H_2O$ synthesis. At higher loadings (> 3 wt. %), formation of promoter-containing bulk phases, such as oxides or phosphates, may also occur. We briefly discuss here the findings of recent studies of promoted VPO catalysts with respect to promoter locations and functions in *n*-butane oxidation to maleic anhydride. The preparation methods and catalytic behavior of promoted VPO catalysts in *n*-butane oxidation to maleic anhydride reported in some of these key studies are summarized in Table 3.

Zazhigalov *et al.* [108] modified the redox properties of the VPO catalysts by incorporating metallic Co in the VPO lattice, which improved desorption of maleic anhydride and enhanced the selectivity to maleic anhydride. Volta *et al.* [109-111] promoted their VPO catalysts with metallic Co and Fe and studied their catalytic behavior in *n*-butane oxidation. They found that both promoters improved the selectivity to maleic anhydride. However, the presence of Fe resulted in increased *n*-butane conversion, while the Co addition led to decreased catalytic activity. They suggested that these promoters affected to a different extent the dispersion of various $VOPO_4$ and $(VO)_2P_2O_7$ phases during precursor activation, which was responsible for the observed behavior. In a different report, Volta *et al.* [112] suggested that Co delayed the transformation of the VPO precursor leading to an enhancement of disorder in the inorganic lattice at lower temperature which stabilized the amorphous VPO phase and led to lower *n*-butane conversion, but higher selectivity to maleic anhydride as compared to the unpromoted VPO catalyst.

Cornaglia *et al.* [113,114] studied the effect of Co cations on the performance of VPO catalysts prepared by impregnation of $VOHPO_4 \cdot 0.5H_2O$ with cobalt acetate and acetyl acetonate (Table 3). They found that Co added at 4 wt. %

Table 3 *Selected promoted VPO catalysts investigated in n-butane oxidation to maleic anhydride*

Promoter	*Loading*	*Synthesis*	*Catalytic performance (Promoted)*	*Catalytic performance (Unpromoted)*	*Promoter location or function*	*Ref.*
Co	4% wt.	Impregnation of VHP* with Co acetate or acac†	‡S(MA) = 60% at ‡C = 80% at 703 K	S(MA) = 30% at C = 80% at 703 K	Co phosphate phase	113, 114
Co	1-5 % at.	Co acac added to V_2O_5, *i*-ButOH and H_3PO_4 under reflux	S(MA) = 71.4% at C = 25% at 673 K (1 at. % Co 1%)	S(MA) = 50.3% at C = 25% at 673 K	Stabilizes amorphous VPO phase	116
Fe	1-5 % at.	Fe acac added to V_2O_5, *i*-ButOH and H_3PO_4 under reflux	S(MA) = 73.5% at C = 25% at 673 K (1% at. Fe)	S(MA) = 50.3% at C = 25% at 673 K	solid solution in $(VO)_2P_2O_7$	116
Ce-Fe	5 % wt.	Mixed oxide precipited from nitrates and added to V_2O_5, *i*-ButOH, benzyl alcohol and H_3PO_4 under reflux	‡Y(MA) = 39-42 % at 673 K.	Y(MA) = 27 % at 673 K.	Most likely as surface Ce-Fe oxide	118
Bi	Bi/V = 0.1 Bi/V = 0.1	Ball-milling of $BiPO_4$ and organic VHP Ball-milling of Bi_2O_3 and organic VHP	S(MA) = 74% at C = 88% at 713K S(MA) = 71% at C = 83% at 713K	S(MA) = 67% at C = 73% at 713K	Bulk $BiPO_4$ Surface $BiPO_4$	124
Mo-Zr-Zn	V/P/Zr/Mo/Zn = 1.0/1.2/0.03/0.03/0.03 (at.)	Zr nitrate, Zn acetate and ammonium molybdate added to V_2O_5, HCl and water. Citric acid then added as a complexing agent.	S(MA) = 88% at C = 46% at 673 K	Not reported	Cations occupying vacant surface and bulk sites in VPO phases	127

Nb	Nb/V = 0.01 (at.)	Organic VHP added to Nb ethoxide in *i*-ButOH under reflux	Y(MA) = 52.5% at 673 K	Y(MA) = 41% at 673 K	Solid solution: $V_{1-x}Nb_xO_2P_2O_{7+x}$	130
Nb, Si. Ti, V Zr	0.25 wt. %	Incipient wetness impregnation of VHP with Nb, Si, Ti, V and Zr alkoxides	S(MA) = 58% at C = 15% for Nb-VPO catalyst at 653 K.	S(MA) = 41% at C = 15% at 653 K	Surface enrichment by XPS; may partially form a solid solution	131

* VHP = $VOHPO_4{\bullet}0.5H_2O$ † acetyl acetonate = acac ‡ C, S(MA) and Y(MA) are, respectively, *n*-butane conversion and the molar selectivity and yield of maleic anhydride [9].

significantly improved the yield of maleic anhydride. The superior catalytic performance of their catalysts was associated with an optimum concentration of very strong Lewis acid sites, very low concentration of isolated V^V centers and the absence of V^V phases. Furthermore, the cobalt phosphate phase detected by XPS aided in stabilizing the excess surface phosphate. The selectivity to maleic anhydride was 60 mol. % for the promoted and 30% for the unpromoted catalyst at 80% *n*-butane conversion at 703 K. Other studies [107,115] suggested that the presence of Co in the VPO phases enhanced the surface enrichment in phosphate and modified the surface acidity increasing the concentration of strong Lewis acid sites.

Hutchings *et al.* [116] studied the effect of Co and Fe ions added during the $VOHPO_4{\cdot}0.5H_2O$ synthesis on *n*-butane oxidation to maleic anhydride (Table 3). At low levels (1-5 at. %), both Co and Fe significantly enhanced the selectivity and intrinsic activity in maleic anhydride formation expressed as moles of maleic anhydride/$m^2 \cdot h$. For instance, the selectivity to maleic anhydride at 25 % *n*-butane conversion was >63 mol. % for the promoted phases and only 50 mol. % for the unpromoted VPO catalyst at 673 K. The authors suggested that Co was insoluble in the $(VO)_2P_2O_7$ phase. Although Co was found to form a solid solution in $VOHPO_4{\cdot}0.5H_2O$, it migrated to a disordered V^{4+}-V^{5+} phosphate phase during thermal activation of $VOHPO_4{\cdot}0.5H_2O$. Therefore, Hutchings *et al.* proposed that Co present at the surface functions as a promoter for the disordered VPO phases indicating that these disordered phases may play an important role in selective oxidation of *n*-butane. On the other hand, Fe ions may be soluble in the $(VO)_2P_2O_7$ lattice and, therefore, function as an electronic promoter for this phase. Satsuma *et al.* [117] reported a novel approach for the introduction of promoters into VPO phases. Lamellar vanadyl benzylphosphate (LVBP) was used as a host material and an Fe complex, $Fe(acac)_3$, as a guest promoter compound. It was found that $Fe(acac)_3$ was successfully inserted into LVBP by heating it in a toluene solution of $Fe(acac)_3$. It was confirmed that $Fe(acac)_3$ was uniformly dispersed in the interlayer space of LVBP without significant disruption of the lamellar structure. However, *n*-butane oxidation over these lamellar VPO phases has not been investigated.

VPO catalysts promoted with a mixed Ce-Fe oxide (5 wt. %) (Table 3) were investigated by Shen *et al.* [118]. They found that the promoted catalyst exhibited higher conversion and selectivity to maleic anhydride than the unpromoted catalyst during *n*-butane oxidation in the absence of gas-phase oxygen. The maleic anhydride yield for the promoted catalysts was much higher (39-42 mol. %) than for the unpromoted VPO catalyst (27 mol. %) at 673 K. The authors suggested that the introduction of Ce-Fe oxides improved the redox performance of VPO catalysts by increasing the lattice oxygen activity. The authors suggested that Ce leads to the formation of highly selective oxygen species, while Fe increases the reactivity of oxygen species.

Zazhigalov *et al.* [119] studied the incorporation of alkali and alkaline-earth metal ions in the VPO lattice. They found that Li, Na, K, Cs, Be, Mg, Ca and Ba cations present at different concentrations easily donated electrons to the VPO lattice with P/V ratios of 1.07 and 1.20, leading to increased negative charge on lattice oxygen atoms and enhancing the rate of *n*-butane oxidation. The presence of these promoters caused an increase of the surface P/V ratio and corresponding changes of the surface acidity. However, the preparation of a catalyst characterized by high activity in *n*-butane oxidation and high selectivity to maleic anhydride still requires fine-tuning of the basicity of surface oxygen atoms to facilitate the activation of *n*-butane and the surface acidity to control the residence times of the reaction intermediates.

Bi is one of the most efficient promoters of VPO catalysts, which was used to improve the selectivity to maleic anhydride in *n*-butane oxidation [106,120,121]. Zazhigalov *et al.* [122,123] found that the incorporation of Bi at Bi/V = 0.1 increased the surface phosphate concentration and enhanced the acidity and the strength of acid sites of the VPO catalysts that improving their catalytic performance in *n*-butane oxidation. In particular, the selectivity to maleic anhydride improved with an increase in the Brønsted acidity. Haber *et al.* [124] further studied the effect of Bi on the catalytic performance of organic VPO catalysts (Table 3). Most notably, Bi-promoted VPO catalysts with significantly enhanced selectivity to maleic anhydride were obtained by mechanochemical treatment of bulk $BiPO_4$ and Bi_2O_3 with $VOHPO_4{\cdot}0.5H_2O$. They observed that Bi_2O_3 formed a surface Bi phosphate phase in the VPO catalysts due to phosphate migration to the Bi_2O_3 phase and that $BiPO_4$ ball-milled together with $VOHPO_4{\cdot}0.5H_2O$ was finely dispersed as a bulk phase in the $BiPO_4{\cdot}(VO)_2P_2O_7$ catalyst. In addition, the mechanochemical treatment produced thinner $VOHPO_4{\cdot}0.5H_2O$ platelets and increased the surface area of promoted catalysts. The presence of the surface Bi phosphate phase in the $Bi_2O_3{\cdot}(VO)_2P_2O_7$ catalyst was claimed to be responsible for the slightly lower selectivity to maleic anhydride displayed by this catalyst (Table 3). Lopez-Nieto *et al.* [125] found that the incorporation of Bi in the VPO catalysts led to a modification in the surface properties that resulted in improved catalytic behavior. They claimed that the incorporation of Bi stabilized the desirable $(VO)_2P_2O_7$ structure, led to an increase in the specific surface area and a decrease in the O 1s-electron binding energy, which enhanced the rate of *n*-butane oxidation. On the other hand, Taufiq-Yap *et al.* [126] found that Bi incorporation into the $(VO)_2P_2O_7$ lattice resulted in the surface area increase as well as the reduction of the V oxidation state (from +4.24 to +4.08). However, no catalytic performance for the Bi-VPO catalyst was reported.

Ji *et al.* [127] promoted VPO catalysts by a mixture of Mo, Zr and Zn ions finding this particular system to be highly selective for the partial oxidation of *n*-butane to maleic anhydride. The combination of these three species created a synergistic effect that enhanced the performance of the promoted VPO catalysts. The selectivity to maleic anhydride of 88 mol. % at 46 % *n*-butane conversion at 673 K was reported for the Mo-Zr-Zn-VPO catalysts. However, no comparison was made with unpromoted VPO catalysts in their study. The authors suggested that these cations were incorporated in the vacant surface sites as well as the bulk VPO lattice. Pt [128], La-Bi [125] and Zr [125] were among other recently investigated VPO promoters. Pt was incorporated as H_2PtCl_6 into the layered $VOHPO_4{\cdot}0.5H_2O$ precursor during its synthesis in aqueous and organic media [128]. However, the catalytic performance of the Pt-VPO catalyst was demonstrated only for the hydrogenation of nitrobenzene and the oxidation of tetrahydrofuran. La-Bi and Zr promoters were introduced as nitrates during organic $VOHPO_4{\cdot}0.5H_2O$ precursor synthesis in butanol [125]. Although the catalytic behavior of the La-Bi and Zr-promoted catalysts in *n*-butane oxidation was not reported, these catalysts were active for the oxidation of ethane to ethylene.

Nb is another highly attractive promoter for the VPO catalysts. Volta *et al.* [129,130] have demonstrated that Nb is an effective promoter of the activity of the VPO catalysts in *n*-butane oxidation to maleic anhydride. Nb increases the number of defects at the surface of $(VO)_2P_2O_7$ platelets, e.g. O-vacancies, and facilitates the initial C-H activation of *n*-butane. These sites mainly correspond to coordinatively unsaturated cations acting as Lewis acid sites of low acidity. Furthermore, Volta *et al.* proposed that the observed improvement in catalytic properties resulted from a redox effect of Nb on the V^{5+}/V^{4+} balance. They proposed that the incorporation of Nb into the VPO framework occurs via substitution of V^{4+} by Nb^{5+} in VO^{2+}, i.e. $(VO)_2P_2O_7 + Nb^{5+} = V_{1-x}Nb_xO_2P_2O_{7+x}$. The observed yield of maleic anhydride was 52.5 mol. % for the Nb-promoted VPO catalyst and only 41 mol. % for the unpromoted catalyst at 673 K. Guliants *et al.* [131] introduced the Nb, Si, Ti, V and Zr promoters at 0.14-0.26 theoretical monolayer surface coverage into model organic $(VO)_2P_2O_7$ catalysts possessing thin platelet crystal morphology that preferentially exposed the surface (200) planes. They found that the promoted $(VO)_2P_2O_7$ catalysts containing surface niobia species were highly active and selective catalysts in *n*-butane oxidation to maleic anhydride. The Nb-promoted $(VO)_2P_2O_7$ catalyst containing only 0.26 monolayer NbO_x surface coverage displayed 58 mol. % selectivity to maleic anhydride at 15 % *n*-butane conversion, while the unpromoted catalyst at the same conversion was characterized by 41 mol. % selectivity at 653 K. The authors proposed that the surface acidity (i.e. Nb species and -POH groups at superstoichiometric P/V ratios) were mainly responsible for the improved catalytic performance of the Nb-promoted $(VO)_2P_2O_7$ catalyst. They further suggested that the promoter elements may partially form a solid solution.

Hutchings *et al.* [132] studied *n*-butane oxidation over $(VO)_2P_2O_7$ catalysts promoted with group 13 elements (B, Al, Ga, In). These promoters were found to change the morphology of the $VOHPO_4{\cdot}0.5H_2O$ precursor from the platelet-like to rosette-like. Although all of these promoters functioned as modifiers of the VPO crystal morphology, only In- and Ga-promoted catalysts displayed improved catalytic performance in *n*-butane oxidation to maleic anhydride. However, the catalytic role of these elements is currently poorly understood.

Aluminum phosphate-promoted VPO catalysts displayed improved catalytic performance during *n*-butane oxidation and shortened equilibration time as compared to conventional VPO catalysts [133]. Aluminum phosphate was incorporated during the synthesis of the $VOHPO_4{\cdot}0.5H_2O$ precursor. The morphology of the $VOHPO_4{\cdot}0.5H_2O$ precursor was influenced by the presence of the aluminium phosphate. The authors [133] found that both $(VO)_2P_2O_7$ and the original amorphous aluminium phosphate crystallized during the equilibration of the promoted $VOHPO_4 \cdot 0.5H_2O$ precursor. However, no mixed V-Al phosphate phase was observed. Moreover, the reasons for the improvement in the catalytic properties of the promoted system were not discussed.

The results of recent studies summarized in Table 3 and briefly discussed in this section demonstrate the beneficial effects of promoting the VPO system for the partial oxidation of *n*-butane to maleic anhydride. However, the specific roles of promoters in modifying the morphology, phase and elemental (bulk and surface) compositions, structures and redox properties of the VPO catalysts at present are poorly understood. Improved fundamental understanding of the VPO promoter effects will enable rational design VPO catalysts with enhanced catalytic performance in *n*-butane oxidation to maleic anhydride. Therefore, detailed studies of several classes of well-defined promoted VPO catalysts containing promoters (1) in solid solution with the VPO lattice, (2) as surface species and (3) nanosized oxides or phosphates, etc., are expected to provide critical fundamental insights into the specific roles of key promoter species in selective oxidation of *n*-butane.

8 New Synthesis Routes to VPO Catalysts

Conventional synthesis methods offer limited control over desirable phase and elemental (bulk and surface) compositions, preferential exposure of active and selective surface planes, surface areas and pore structures of VPO catalysts, which define their catalytic performance in selective oxidation of *n*-butane. For instance, conventional methods produce VPO catalysts with relatively low surface areas (<20 m^2/g) and a limited choice of crystal morphologies.

Recently reported meso- and macroscale self-assembly approaches conducted, respectively, in the presence of surfactant mesophases [134-136] and colloidal sphere arrays [137] are highly promising for the *molecular engineering* of novel catalytic mixed metal oxides. These novel methods offer the possibility to control surface and bulk chemistry (e.g. the V oxidation state and P/V ratios), wall nature (i.e. amorphous or nanocrystalline), morphology, pore structures and surface areas of mixed metal oxides. Furthermore, these novel catalysts represent well-defined model systems that are expected to lead to new insights into the nature of the active and selective surface sites and the mechanism of *n*-butane oxidation. In this section, we describe several promising synthesis approaches to VPO catalysts, such as the self-assembly of mesostructured VPO phases, the synthesis of macroporous VPO phases, intercalation and pillaring of layered VPO phases and other methods.

8.1 Mesostructured VPO Phases. – The mesoscale self-assembly pathway is illustrated in Figure 17 for mesostructured metal oxides possessing a 2D-hexagonal pore structure. This general route was initially proposed for mesostructured silicate phases. The formation of highly ordered mesostructures containing cylindrical mesopores takes place via charge-matching interactions between the inorganic species at the interface with a pre-existing hexagonal lyotropic liquid

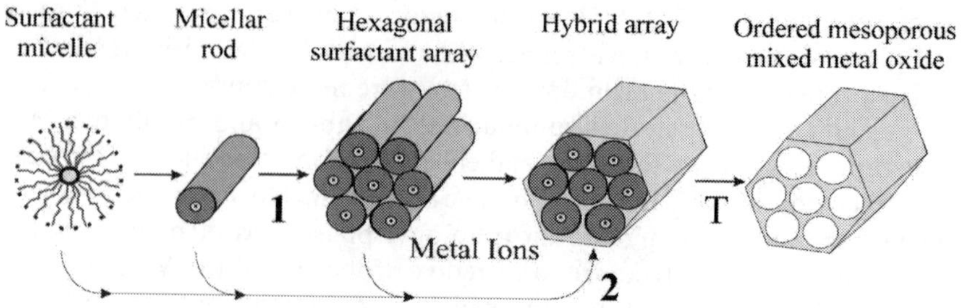

Figure 17 *Formation of 2D-hexagonal* mesoporous *mixed metal oxides by liquid crystal templating mechanism [134,135]*

crystal phase. Then, the template removal leads to highly ordered mesoporous structures [134-136].

Mesostructuring of VPO phases offers the possibility to design novel VPO catalysts with high surface areas, controlled porosities, tunable surface compositions and improved catalytic performance. Few reports exist on the synthesis of mesostructured VPO phases. Iwamoto *et al.* [138] reported the synthesis of mesostructured hexagonal vanadium-phosphorus oxide phases using alkyltrimethyl ammonium surfactants. However, the mesostructure in these phases was lost upon calcination. Doi and Miyake [139] reported the synthesis of a hexagonal mesostructured VPO compound from $VOHPO_4{\cdot}0.5H_2O$ by surfactant intercalation and subsequent hydrothermal treatment. However, these materials suffered from poor thermal stability and low phosphorus content that are detrimental for their potential applications in *n*-butane oxidation. Amoros *et al.* [140] described the synthesis of hexagonal mesostructured oxovanadium phosphates. However, the surfactant removal from their vanadium phosphates resulted in the collapse of the mesostructure. Mizuno *et al.* [141] described the synthesis of hexagonal, cubic and lamellar mesostructured vanadium-phosphorus-oxides that lost the structural order upon thermal treatment. Datta *et al.* [142] synthesized mesostructured lamellar VPO phase employing hexadecylamine as a template. However, no attempts were made to remove the surfactant.

Recently, Carreon and Guliants reported novel hexagonal, cubic and lamellar VPO phases, which displayed improved thermal stability, desirable chemistries (i.e. the P/V ratios and vanadium oxidation states), and pore structures for the partial oxidation of *n*-butane [143-145]. These novel VPO phases displayed the selectivities to maleic anhydride up to ~ 40 mol. % at 673K at ~ 10 % *n*-butane conversion [146]. A conventional organic VPO catalyst containing well-crystallized vanadyl(IV) pyrophosphate, the proposed active and selective phase for *n*-butane oxidation to maleic anhydride, displayed the selectivities to maleic anhydride 50 mol. % under the same reaction conditions. The low yields observed for mesoporous VPO catalysts confirmed the critical role of the vanadyl pyrophosphate phase $(VO)_2P_2O_7$ in catalyzing the oxidation of *n*-butane to maleic anhydride. Therefore, the amorphous nature of the mesoporous VPO samples may be responsible for the observed poor maleic anhydride yields. On the other hand, the mesoporous VPO phases were thermally unstable under n-butane oxidation conditions. The BET surface area measurements of the fresh and used catalysts indicated the collapse of the mesoporous structures with concomitant decrease of the surface area from ~ 190 to 15 m^2/g. Low n-butane conversions observed at even short reaction times (<1 hr) suggested that the mesoporous structure collapsed very quickly under reaction conditions. Furthermore, the XRD diffraction pattern of the used catalyst showed no reflections at low 2θ angles confirming the mesostructure collapse. Only dense VPO phases such as γ-$VOPO_4$ were observed. The incomplete cross-linking of the mesoporous VPO framework, its low Tammann temperature and redox properties are mainly responsible for its poor thermal stability.

8.2 Macroporous VPO Phases. – The macroscale templating of bulk mixed metal oxide phases in the presence of colloidal sphere arrays typically consists of three steps shown in Figure 18. First, the interstitial voids of the monodisperse sphere arrays are filled with metal oxide precursors. In the second step, the precursors condense and form a solid framework around the spheres. Finally, the spheres are removed by either calcination or solvent extraction leading to the formation of 3D ordered macroporous structures [137].

Macroporous VPO phases with tunable phase compositions, pore structures and morphologies have been recently synthesized by employing monodisperse polystyrene sphere arrays as templates [147,148]. Colloidal polystyrene spheres were ordered into closed-packed arrays by sedimentation and centrifugation. Depending on the choice of VPO sources and template removal method, various crystalline VPO phases were obtained, e.g. $(VO)_2P_2O_7$, γ-$VOPO_4$, $VOHPO_4{\cdot}0.5H_2O$. The macroscale-templated synthesis produced VPO phases with unprecedented high surface areas (75 m^2/g), desirable macroporous structures, optimal bulk compositions ($P/V \sim 1.1$), desirable vanadium oxidation

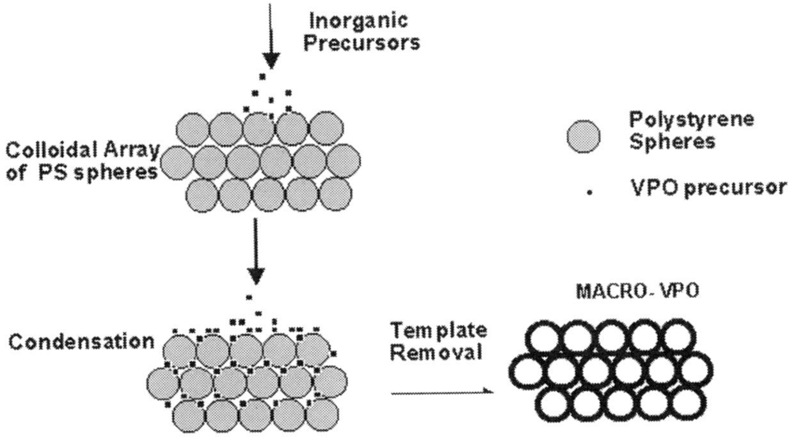

Figure 18 *Synthesis of highly ordered* macroporous *VPO phases [149]*

states (+4.1(184)4.4) and preferential exposure of the surface (200) planes of $VO_2P_2O_7$, the proposed active and selective phasefor the partial oxidation of *n*-butane. The catalytic performance of macroporous VPO catalysts was evaluated in the partial oxidation of *n*-butane to maleic anhydride [149]. The observed yield of maleic anhydride was > 50 mol. % for the macroporous VPO catalyst that displayed ordered arrays of 380 nm pores and walls composed of 20 nm crystallites of $(VO)_2P_2O_7$. Under similar reaction conditions, the yield of maleic anhydride was ~ 40 mol. % for a conventional organic VPO catalyst. The ordered open pore structures composed of nanocrystalline $(VO)_2P_2O_7$ walls and high surface areas (> 40 m^2/g) in macroporous VPO catalysts enhanced the access of reactant molecules to the active and selective surface which led to improved catalytic performance. These ordered macroporous VPO phases are highly promising for potential applications in selective catalytic oxidation, representing model catalytic systems that will enable improved fundamental understanding of the relationships between the molecular structure and catalytic properties of a broad range of industrially relevant catalytic systems.

8.3 Intercalation and Pillaring of Layered VPO Phases. – Guliants *et al.* reported new precursors to vanadyl(IV) pyrophosphate catalysts for *n*-butane oxidation based on the structural modification of the layered $VOHPO_4{\cdot}0.5H_2O$ precursor [150-153]. The $VOHPO_4{\cdot}0.5H_2O$ precursor was intercalated with *n*-alkylamines in neat amine which resulted in a significant expansion of the interlayer spacing as well as disordered stacking of the vanadyl(IV) hydrogen phosphate layers. However, this structural disorder in the intercalated precursor phases led to an improved catalytic performance of the VPO catalysts in *n*-butane oxidation after their thermal transformation [150,153]. Vanadyl(IV) phosphite and alkylphosphonates containing vanadyl(IV) dimers in their layered structures transformed into vanadyl(IV) pyrophosphate catalysts with higher surface areas (35 m^2/g) and higher selectivity to maleic anhydride than conventional organic VPO catalysts [151,152]. Furthermore, these new VPO precursor phases transformed into vanadyl(IV) pyrophosphate catalysts at lower temperatures than the conventional organic precursor (i.e. 550-650 K). Thermally stable pillared VPO catalyst with 17.0 (199)(186) interlayer spacing was also reported, which was obtained by the intercalation of mixed vanadyl(IV) phosphite-phosphate phases with aminopropyltriethoxysilane (APS) [153]. These intercalated phases were synthesized by contacting the hemihydrate VPO precursor with a two-fold excess of 0.5 M APS in anhydrous ethanol at room temperature for 48 h. The pillared phase was obtained by calcining the APS-intercalated phase in nitrogen at 673 K. The catalytic performance of the pillared VPO catalyst was reported for *n*-butane oxidation to maleic anhydride at 673 K [153]. The steady-state selectivity to maleic anhydride of 25 mol. % at 45 % *n*-butane conversion was achieved after 6 days in stream. The low selectivity to maleic anhydride observed for the pillared VPO catalyst confirmed that the presence of $(VO)_2P_2O_7$ is critical for selective *n*-butane oxidation to maleic anhydride.

8.4 Alternative Synthesis Methods of Dense VPO Phases. – Other novel synthesis approaches to VPO phases have been recently reported. For instance, Hutchings *et al.* [154,155] have reported a new synthesis route for the VPO catalysts using supercritical CO_2 as an antisolvent. In this capacity, CO_2 can be liquid at <42 bar at 293 K or a supercritical fluid (T_c=304K, P_c=72 bar). This route produced amorphous microspheroidal VPO phases, which were more active (intrinsic activity expressed as moles of maleic anhydride produced per 1 m^2 of catalyst per h) than crystalline $(VO)_2P_2O_7$ catalyst in *n*-butane oxidation to maleic anhydride. Furthermore, these novel phases did not require an extensive activation period to achieve full catalytic activity. These observations, however, contradict a number of published studies that reported that the presence of crystalline vanadyl(IV) pyrophosphate was critical for superior activity and selectivity in *n*-butane oxidation [1-4, 156].

The VPO catalysts consisting of dominant $(VO)_2P_2O_7$ and $VOPO_4$ impurity phases were prepared using ultrasound and microwave irradiation methods [157]. The ultrasound method produced a catalyst in a relatively short time that had similar composition and surface area to those prepared by conventional organic method. In this method, V_2O_5 and H_3PO_4 were mixed with 2-butanol and benzyl alcohol immersed in a water bath and the mixture was subjected to ultrasound irradiation. On the other hand, the VPO catalyst obtained in a microwave synthesis (same precursors and solvents as in the ultrasound method) showed different surface morphology as compared to the products of conventional synthesis. However, the catalytic performance of these novel VPO phases in *n*-butane oxidation has not been reported.

Recently, relatively high surface area VPO catalysts have been prepared in the presence of polyethylene glycols (PEG2000-PEG20000) [158]. The resultant catalysts possessed the specific surface areas of 41-54 m^2/g, with an average particle size of 100-200 nm and crystalline $(VO)_2P_2O_7$ phase, displaying improved catalytic performance for the partial oxidation of *n*-butane to maleic anhydride. For instance, the yields of maleic anhydride > 65 mol. % were observed for the VPO catalysts synthesized in the presence of PEG10000 and PEG20000, while the VPO catalysts obtained in the absence of PEG displayed maleic anhydride yields <55 mol. % under similar reaction conditions. Most likely the PEG acts as a pore former increasing the specific surface area of the resultant VPO phases.

In the same study [158] high surface area VPO catalysts were prepared by a modified mechanochemical route. This method consisted of three distinct steps: (i) synthesis of the $VOHPO_4{\cdot}0.5H_2O$ precursor by controlled addition of phosphoric acid, (ii) pre-conditiong of the $VOHPO_4{\cdot}0.5H_2O$ precursor in nitrogen followed by (iii) ball-milling in cyclohexane [158]. The ball-milling process not only increased the surface area of the VPO catalysts from 15 to 38 m^2/g, but also changed the local environment and the redox behavior of the active surface sites, which resulted in enhanced performance. The ball-milled catalysts displayed the selectivity to maleic anhydride of 59 mol. % at 93 % *n*-butane conversion at 698 K. Similar selectivity was observed for conventional organic VPO catalysts at somewhat lower *n*-butane conversion (83 %). The authors suggested that the ball-milling treatment led to a higher reactivity of the lattice oxygen associated with the V^{5+} and V^{4+} species. The enhancement of catalytic performance was attributed in this study to the increased exposure of the (200) planes of $(VO)_2P_2O_7$ as a direct result of ball-milling, which resulted in fracturing of $(VO)_2P_2O_7$ platelet crystals into thinner platelets.

9 Concluding Remarks

Almost 40 years have passed since Bergman and Frisch reported that VPO catalysts selectively oxidized *n*-butane to maleic anhydride. The *n*-butane oxidation to maleic anhydride is still the only industrial process of selective vapor-phase oxidation of a light alkane. The VPO system has been intensively studied over the last two decades and significant fundamental insights into the mechanism of *n*-butane oxidation to maleic anhydride have been obtained. Crystalline $(VO)_2P_2O_7$ characterized by the preferential exposure of the surface (200) planes and a slight excess of phosphate ($P/V = 1.01$-1.10) is considered as the most active and selective phase for *n*-butane oxidation. The catalytic performance of VPO catalysts in this reaction can be improved by the introduction of certain promoters. However, there is still a lack of agreement in the literature concerning specific roles of these promoters in controlling the morphology, phase and elemental (bulk and surface) compositions, structures and redox properties of VPO catalysts, which define their catalytic performance.

Furthermore, several novel synthesis approaches have recently shown promise for the molecular engineering of new well-defined VPO and other mixed metal oxide catalysts that will enable improved fundamental understanding of the relationships between the molecular structure and catalytic properties of a broad range of industrially relevant catalytic systems. Finely tuned pore sizes, metal oxide wall structures and compositions, high surface areas and enhanced accessibility of the active and selective surface sites in these novel phases are highly attractive for improving the catalytic performance in oxidation of light alkanes. Future studies are expected to focus on the use of these model catalysts to unravel remaining mysteries about the nature of the active and selective VPO phase and surface sites, the mechanism of *n*-butane oxidation and the role of promoters. These insights are critical for the development of molecular surface structure-activity and selectivity relationships and for the establishment of design rules of improved selective oxidation catalysts for light alkane oxidation.

References

1. D.J. Hucknall, *Selective Oxidation of Hydrocarbons*, Academic Press, London 1974.
2. S.T. Oyama, A.N. Desikan, J.W. Hightower, *Catalytic Selective Oxidation*, ACS Symposium Series 523, American Chemical Society: Washington, DC, 1993.
3. J. Haber, *Stud. Surf. Sci. Catal.* 1997, **110**, 1.
4. G. Centi, F. Cavani, F. Trifirò, *Selective Oxidation by Heterogeneous Catalysis. Fundamental and Applied Catalysis*, Kluwer Academic/Plenum Publishers: New

York, 2001.

5. F. Cavani, F. Trifirò, *Appl. Catal.* 1992, **88**, 115.
6. *Encyclopedia of Chemical Technology*, J. I. Kroschwitz, M. Howe-Grant, Eds., New York: Wiley, 1991.
7. R. I. Bergman, N.W. Frisch, US Patent 1966, 3, 293, 268.
8. S. Irvin-Monshaw, A. Klein, *Chem. Eng.* 1989, **96,** 35.
9. *Conversion* (C) of *n*-butane is defined as the ratio of the moles of *n*-butane consumed to moles of *n*-butane fed; the *selectivity* to maleic anhydride, S(MA), is defined as the ratio of the moles of *n*-butane transformed to maleic anhydride to total moles of *n*-butane consumed, and the *yield* of maleic anhydride, Y(MA), is a product of the two quantities, Y(MA) = C x S(MA). *Space velocity* is equal to the volume of the *n*-butane/air feed that passed through one reactor volume in 1 hour.
10. G. Centi, F. Trifirò, J.R. Ebner, V.M. Franchetti, *Chem. Rev.* 1988, **88,** 55, and refs. therein.
11. P. Ruiz, B. Delmon, *Catal. Today* 1987, **1**,2, and refs. therein.
12. G. Centi, *Catal. Today*1993, **16,** 1, and refs. therein.
13. F. Cavani, F. Trifirò, *Catalysis*, 1994, **11,** 246, and refs. therein.
14. J. R. Ebner, M.R. Thompson, In *Structure-Activity and Selectivity Relationships in Heterogeneous Catalysis*, R.K. Grasselli, A.W. Sleight, Eds., Elsevier, Amsterdam, 1991, 31.
15. R. M. Contractor, J.R. Ebner, M.J. Mummy, In *New Developments in Selective Oxidation*, G. Centi, F.Trifirò, Eds., Elsevier, Amsterdam, 1990, 553.
16. H. S. Horowitz, C.M. Blackstone, A.W. Sleight, G. Teufer, *Appl. Catal.* 1988, **38,** 193.
17. I. Matsuura, *Catal. Today*, 1993, **16,** 61.
18. T. Okuhara, M. Misono, *Catal. Today* 1993, **16,** 61.
19. L. M. Cornaglia, C.A. Sanchez, E.A. Lombardo, *Appl. Catal.* 1993, **95,** 117.
20. E. Bordes, *Catal. Today* 1993, **16,** 27.
21. D. Ye, A. Satsuma, A. Hattori, T. Hattori, Y. Murakami, *Catal. Today* 1993, **16,** 113.
22. T. Shimoda, T. Okuhara, M. Misono, *Bull. Chem. Soc. Jpn.* 1985, **58,** 2163.
23. L.M. Cornaglia, C. Caspani, E.A. Lombardo, *Appl. Catal.* 1991, **74,** 15.
24. W.A. Waters, J.S. Littler, In *Oxidation in Organic Chemistry*; K.B. Wiberg, Ed.; Academic Press: New York, 1965; Part A, 185;
25. R.H. Holm, *Chem. Rev.* 1987, **87,** 1401;
26. A. Butler, G.E. Meister, M.J. Clague, *Chem. Rev.* 1994, **94,** 625.
27. J. Rocek, D.E. Aylward, *J. Am. Chem. Soc.* 1975, **97,** 5452.
28. F. Hasan, J. Rocek, *J. Am. Chem. Soc.* 1974, **96,** 534.
29. J. Rocek, A.E. Radkowsky, *J. Am. Chem. Soc.* 1973, **95,** 7123.
30. L.P. Hammett, *Physical Organic Chemistry*, McGraw-Hill: New York, 1970, Ch. 9.
31. C.F. Baes, R.E. Mesmer, In *The Hydrolysis of Cations*; Wiley: New York, 1976; 197.
32. M.J. Gresser, A.S. Tracey, In *Vanadium in Biological Systems: Physiology and Biology*, N.D. Chasteen, Ed., Kluwer Academic Publishers: Boston, 1990, 63.
33. L. Pettersson, I. Andersson, B. Hedman,*Chem. Scr.* 1985, **25,** 309.
34. A.A. Ivakin, L.D. Kurbatova, M.V. Kruchinina, *Russ. J. Inorg. Chem.* 1986, **31,** 219.
35. A.S. Tracey, B. Galeffi, S. Mahjour, *Can. J. Chem.* 1988, **66,** 2294.
36. B. Galeffi, A.S. Tracey, *Can. J. Chem.* 1988, **66**, 2565.
37. A.S. Tracey, M.J. Gresser, *Can. J. Chem.* 1988, **66**, 2570.
38. A.S. Tracey, M.J. Gresser, B. Galeffi, *Inorg. Chem.* 1988, **27**, 157.
39. M.J. Gresser, A.S. Tracey, K.M. Parkinson, *J. Am. Chem. Soc.* 1986, **108**, 6229.
40. L. de Meis, *J. Biol. Chem.* 1984, **259,** 6090.

41. M.M. Craig, M.J. Gresser, *J. Cell. Biol.* 1989, **107**, 189a.
42. V.V. Guliants, S.A. Holmes, J.B. Benziger, P. Heaney, D. Yates, I.E. Wachs, *J. Mol. Catal. A*2001, **172**, 265.
43. E. Bordes, *Catal. Today* 1987, **1**, 499, and refs. therein.
44. Z.-Y. Xue, G.L. Schrader, *J.Phys. Chem. B.* 1999, **103**, 9459.
45. E. Bordes, P. Courtine, J.W. Johnson, *J. Solid Sate Chem.* 1984, **55**, 270.
46. M.R. Thompson, A.C. Hess, J.C. White, J. Anchell, J.B. Nicholas, M.I. McCarthy, J.R. Ebner, F.W. Lytle, In *Preprints II World Congress and IV European Workshop Meeting on New Developments in Selective Oxidation*, V. Cortes Corberan, S.Vic Bellon, Eds.; 1993, C1.
47. C.C. Torardi, Z.G. Li, H.S. Horowitz, W. Liang, M.-H. Whangbo, *J. Solid State Chem.* 1995, **119, 2,** 349.
48. S. Albonetti, F. Cavani, F. Trifiro, P. Venturoli, G. Calestani, M. Granados, J.L.G. Lopez; Fierro, *J. Catal.* 1996, **160, 1**, 52.
49. H. Morishige, J. Tamaki, N. Miura, N. Yamazoe, *Chem. Lett.*, 1990, 1513.
50. N. Yamazoe, H. Morishige, J. Tamaki, N. Miura, In *New Frontiers in Catalysis*, L. Guczi, *et al.* Eds., Elsevier: Amsterdam, 1993, 1979.
51. H. Igarashi, K. Tsuji, T. Okuhara, M. Misono, *J. Phys. Chem.* 1993, **97,** 7065.
52. E.A. Lombardo, C.A. Sanchez, L.M. Cornaglia, *Catal. Today* 1992, **15,** 407.
53. G.A. Sola, B.T. Pierini, J.O. Petunchi, *Catal. Today*. 1992, **15,** 537.
54. P. Gai, K. Kourtakis, *Science* 1995, **267,5198**, 661.
55. V.V. Guliants, J.B. Benziger, S. Sundaresan, N. Yao, I.E. Wachs, *Catal. Lett.* 1995, **32**, 379.
56. I. Matsuura, M. Yamazaki, *In New Developments in Selective Oxidation*, G. Centi, F. Trifirò, Eds.; Elsevier: Amsterdam, 1990, 563.
57. M. O'Connor, F. Dason, B.K. Hodnett, *Appl. Catal.* 1990, **64,** 161.
58. J.R. Ebner, M.R. Thompson, *Catal. Today* 1993, **16**, 51.
59. M. Guilhoume, M. Roullet, G. Pajonk, B. Grzybowska, J.C. Volta, In *New Developments in Selective Oxidation by Heterogeneous Catalysis*, P. Ruiz, B. Delmon, Eds.; Elsevier: Amsterdam, 1992, 255.
60. F. Cavani, G. Centi, F. Trifirò, *La Chimica & L'Industria*, Milan, 1992, **74,** 182.
61. G. Bergeret, M. David, J.P. Broyer, J.C. Volta, G. Hecquet, *Catal. Today* 1987, **1,** 37, and refs. therein.
62. N. Harrouch-Batis, H. Batis, A. Ghorbel, J.C. Vedrine, J.C. Volta, *J. Catal.* 1991, **128,** 248.
63. V.V. Guliants, J.B. Benziger, S. Sundaresan, I.E. Wachs, J.M. Jehng, J.E. Roberts, *Catal. Today* 1996, **28**, 275.
64. G.J. Hutchings, C.J. Kiely, M.T. Sananes-Schulz, A. Burrows, J.C. Volta, *Catal. Today* 1998, **40(2-3)**, 273.
65. M.T. Sananes-Schulz, A. Tuel, G.J. Hutchings, J.C. Volta, *J. Catal.* 1997, **166,2,** 388
66. M. Misono, K. Miyamoto, K. Tsuji, T. Goto, N. Mizuno, T. Okuhara, In *New Developments in Selective Oxidation by Heterogeneous Catalysis*, G. Centi, F. Trifirò, Eds.; Elsevier: Amsterdam, 1990, 605.
67. I. Matsuura, M. Yamazaki, M. *In New Developments in Selective Oxidation*, G. Centi, F. Trifirò, Eds.; Elsevier: Amsterdam, 1990, 563.
68. J.C. Volta, K. Bere, Y.J. Zhang, R. Olier, In *Catalytic Selective Oxidation*, J. Hightower, T. Oyama, Eds.; ACS: Washingon, DC, 1993, 217.
69. T. Okuhara, K. Inumaru, M. Misono, In *Catalytic Selective Oxidation*, J. Hightower, T. Oyama, Eds.; ACS: Washingon, DC, 1993, 156.
70. J.T. Gleaves, J.R. Ebner, T.C. Kuechler, *Catal. Rev. – Sci. Eng.* 1988, **30,** 49.

71. J. March, *Advanced Organic Chemistry*, 4th Ed.; Wiley: New York, 1992.
72. F. Cavani, F. Trifirò, *Appl. Catal. A*, 1997, **157, 1-2,** 195.
73. S.O. Shekhtman, G.S. Yablonsky, S. Chen, J.T. Gleaves, *Chem. Eng. Science* 1999, **54, 20**, 4371.
74. Z.-Y Xue, G.L. Schrader, *J. Catal.* 1999, **184, 1**, 87.
75. P.A. Agaskar, L. DeCaul, R.K. Grasselli, *Catal. Lett.* 1994, **23,** 339.
76. S.J. Puttock, C.H. Rochester, *J. Chem. Soc., Faraday Trans. 1* 1986, **82**, 2773.
77. S.J. Puttock, C.H. Rochester, *J. Chem. Soc., Faraday Trans. 1* 1986, **82,** 3013.
78. B. Chen, E.J. Munson, *J. Am. Chem. Soc.*,1999, **121**, 11024.
79. B. Chen, E.J. Munson, *J. Am. Chem. Soc.*,2002, **124, 8**, 1638.
80. M.A. Pepera, J.L. Callahan, M.J. Desmond, E.C. Milberger, P.R. Blum, N.J. Bremer, *J. Am. Chem. Soc.* 1985, **107**, 4883.
81. D. Wang, M. Barteau, *J. Catal.* 2001, **197**, 17.
82. D. Wang, M. Barteau, *Appl. Catal. A*, 2002, **223**, 205.
83. D. Wang, M. Barteau, *Catal. Lett.* 2003, **90, 1-2**, 7.
84. B. Bachiller-Baeza, I. Rodriguez-Ramos, J.C. Volta, M. Cerro-Alarcon, A. Guerrero Ruiz, *Stud. Surf. Sci. Catal.*2001, **138**, 379.
85. H.H. Kung, *Ind. Eng. Chem. Prod. Res. Dev.* 1986, **25**, 171.
86. H.F. Liu, R.-S. Liu, K.Y. Liew, R.E. Johnson, J.H. Lunsford, *J. Am. Chem. Soc.* 1984, **106**, 4117.
87. R. Atkinson, *Chem. Rev.* 1985, **85**, 69.
88. Y.-M. Chen, D.E. Clemmer, P.B. Armentrout, *J. Chem. Phys.* 1993, **98**, 4929.
89. P.B. Armentrout, J.L. Beauchamp, *Acc. Chem. Res.* 1989, **22,** 315.
90. J.A.M. Simões, J.L. Beauchamp, *Chem. Rev.* 1990, **90**, 629.
91. E.R. Fisher, P.B. Armentrout, *J. Am. Chem. Soc.* 1992, **114**, 2049.
92. R.G. Wilkins, In *Kinetics and Mechanism of Reactions of Transition Metal Complexes*, 2nd Thoroughly Revised Edition, VCH Publishers: New York, 1991.
93. S.W. Benson, *Thermochemical Kinetics*, 2nd Ed., A Wiley-Interscience, New York, 1976.
94. J.E. Huheey, E.A. Keiter, R.C. Keiter, *Inorganic Chemistry. Principles of Structure and Reactivity*, 4th Ed., HarperCollins College Publishers: New York, 1993, A-21.
95. D.E. Clemmer, N. Aristov, P.B. Armentrout, *J. Phys. Chem.* 1993, **97,** 544.
96. G. Centi, F. Trifirò, G. Busca, J.R. Ebner, J. Gleaves, *Faraday Disc. Chem. Soc.* 1989, **87,** 215.
97. V.A. Zazhigalov, J. Haber, J. Stoch, A.I. Pyatnitskaya, G.A. Komashko, V.M. Belousov, *Appl. Catal. A* 1993, **96,** 135.
98. B. Schiøtt, K.A. Jørgensen, *Catal. Today* 1993, **16**, 79, and refs. therein.
99. R.A. Sheldon, J.K. Kochi, *Metal Catalyzed Oxidations of Organic Compounds*, Academic Press: New York, 1981.
100. R.H. Holm, *Chem. Rev.* 1987, **87**, 1401.
101. A. Butler, G.E. Meister, M.J. Clague, *Chem. Rev.* 1994, **94,** 625, and refs. therein.
102. R. Grasselli, In *Surface Properties and Catalysis by Non-Metals*; J. Bonnelle, B. Delmon, E. Derouane, Eds.; Reidel: Dordrecht, The Netherlands, 1983, 273.
103. V.V. Guliants, S.A. Holmes, *J. Mol. Catal. A* 2001, **175, 1-2**, 227.
104. I.E. Wachs, J.-M. Jehng, G. Deo, B.M. Weckhuysen, V.V. Guliants, J.B. Benziger, S. Sundaresan, *J. Catal.* 1997, **170, 1**, 75
105. V.V. Guliants, *Catal. Today* 1999, 51, 2, 255.
106. G.J. Hutchings, *Appl. Catal.* 1991, **72**, 1.
107. G.J. Hutchings, R. Higgins, *J. Catal.* 1996, **162,2,** 153.
108. V.A. Zazhigalov, J. Haber, J. Stoch, A.I. Pyatnitzkaya, G.A. Komashko, V.M.

Belousov, *Appl. Catal. A* 1993, **96**, 135.
109. F. Ben Abdelouahab, R. Olier, M. Ziyad, J.C. Volta, *J. Catal.* 1995, **157**, 687.
110. M.T. Sananes-Schulz, F. Ben Abdelouahab, G.J. Hutchings, J.C. Volta, *J. Catal.* 1996, **163**, 346.
111. V. Martin, J.M.M. Millet, J.C. Volta, *J. Thermal Analysis and Calorimetry* 1998, **53,1,** 111.
112. S. Sajip, J.K. Bartley, A. Burrows, C. Rhodes, J.C. Volta, C.J. Kiely, G.J. Hutchings, *Phys. Chem. Phys.* 2001, **3**, 2143.
113. L. Cornaglia, S. Irusta, E.A. Lombardo, M.C. Durupty, J.C. Volta, *Catal. Today* 2003, **78, 1-4,** 291
114. L.M. Cornaglia, C.R. Carrara, J.O. Petunchi, E.A. Lombardo, *Catal. Today* 2000, **57, 3-4,** 313.
115. G.J. Hutchings, C. Kelly, M.T. Sananes-Schulz, A. Burrows, J.C. Volta, *Catal. Today* 1998,**40**, 273.
116. S. Sajip, J.K. Bartley, A. Burrows, M.T. Sananes-Schulz, A. Tuel, J.C. Volta, C.J. Kiely, G.J. Hutchings, *New J. Chem.* 2001, **25,1,** 125.
117. A. Satsuma, Y. Kijima, S. Komai, Y. Kamiya, E. Nishikawa, T. Hatttori, *Catal. Today* 2001, **71, 1-2,** 161
118. S. Shen, J. Zhou, F. Zhang, L. Zhou, R. Li, *Catal. Today* 2002, **74**, 37.
119. V.A. Zazhigalov, J. Haber, J. Stoch, I.V. Bacherikova, G.A. Komashko, A.I. Pyatnitskaya, *Appl. Catal. A* 1996, **134,2,** 225.
120. N. Yamazoe H. Morishige, J. Tamaki, N. Miura, *Stud. Surf. Sci. Catal.* 1993, **75**, 1989.
121. P. Ruiz, Ph. Bastians, L. Caussin, R. Reuse, L. Dasa, D. Acosta, B. Delmon, *Catal. Today*, 1993, **16,** 99.
122. V.A. Zazhigalov, J. Haber, J. Stoch, E.V. Cheburakova, *Catal. Commun.* 2001, **2**, 375.
123. V.A. Zazhigalov, *Kinetics and Catal.*, 2002, **43, 4,** 514.
124. J. Haber, V.A. Zazhigalov, J. Stoch, L.V. Bogutskay, I.V. Batcherikova, *Catal. Today* 1997,**33**, 39.
125. Solsona, B.; Zazhigalov, V.A.; Lopez Nieto, J.M.; Bacherikova, I.V.; Diyuk, E.A. *Appl. Catal. A*, 2003, **249**, 81.
126. Y.H. Taufiq-Yap, K.P. Tan, K.C. Waugh, M.Z. Hussein, I. Ramli, M.B. Abdul Rahman, *Catal. Lett.* 2003, **89**, 87.
127. L. Xu, X. Chen, W. Ji, Q. Yan, Y. Chen, *React. Kinetics Catal. Lett.*, 2002, **76,2**, 335.
128. A. Datta, M. Agarwal, S. Dasgupta, R.Y. Kelkar, *J. Mol. Catal. A* 2003,**198**, 205.
129. P.G. Pries de Oliveira, J.G. Eon, M. Chavant, A.S. Riche, V. Martin, S. Caldarelli, J.C. Volta, *Catal. Today* 2000, **57, 3-4**, 177.
130. A.M. Duarte-Farias, W.A. Gonzalez, P.G. Oliveira, J-G. Eon, J-M. Herrmann, M. Aouine, S. Loridant, J.C. Volta, *J. Catal.* 2002, **208**, 238.
131. V.V. Guliants, J.B. Benziger, S. Sundaresan, I.E. Wachs, A.M. Hirt, *Catal. Lett.* 1999, **62, 2-4**, 87.
132. J.C. Sanchez, J.A. Lopez-Sanchez, R.P.K. Wells, C. Rhodes, G.J. Hutchings, *J. New J. Chem.* 2001, **25,12,** 1528.
133. S. Holmes, L. Sartoni, A. Burrows, V. Martin, G.J. Hutchings, C. Kiely, J.C. Volta, *Stud. Surf. Sci. Catal.* 2000, **130B**, 1709.
134. C.T. Kresge, M.E. Leonowicz, W.J. Roth, J.C. Vartuli, J.S. Beck, *Nature* 1992, **359**, 710.
135. J.S. Beck, J.C. Vartuli, W.J. Roth, M.E. Leonowicz, C.T. Kresge, K.D. Schmitt, C.T.-W. Chu, D.H. Olson, E.W. Sheppard, S.B. McCullen, J.B. Higgins, J.L.

Schlenker, *J. Am. Chem. Soc.* 1992, **114**, 10834.
136. C.T. Kresge, M.E. Leonowicz, W.J. Roth, J.C. Vartulli (Mobil Oil Corp.) US Patent 1992, 5098684.
137. A. Stein, R.C. Schroden, *Curr. Opin. Solid State Mater. Sci.* 2001, 5, 553.
138. T. Abe, A. Taguchi, M. Iwamoto, *Chem. Mater.* 1995, **7**, 1429.
139. T.Doi, T. Miyake, *Chem. Commun.* 1996, 1635.
140. J.E. Haskouri, M. Roca, S. Cabrera, J. Alamo, A. Beltran-Porter, D. Beltran-Porter, M.D. Marco, P. Amoros, *Chem. Mater.* 1999, **11**, 1446.
141. N. Mizuno, H. Hatayama, S. Uchida, A. Taguchi, *Chem. Mater.* 2001, **13**, 179.
142. S. Dasgupta, M. Agarwal, A. Datta, *J. Mater. Chem.* 2002, **12,** 162.
143. M.A. Carreon, V.V. Guliants, *Microp. Mesopor. Mater.*2002, **55, 3**, 297.
144. M.A. Carreon, V.V. Guliants, *Stud. Surf. Sci. Catal.* 2002, **141**, 301 .
145. M.A. Carreon, V.V.Guliants, *Catal. Today* 2003, **78**, 303.
146. M.A. Carreon, V.V.Guliants, F. Pierelli, F. Cavani, *Catal. Lett.* 2004, **92, 1-2**, 11
147. M.A. Carreon, V.V. Guliants, *Chem. Commun.* 2001, 1438.
148. M.A. Carreon, V.V. Guliants, *Chem. Mater.* 2002, **14, 6,** 2670.
149. M.A. Carreon, PhD. Thesis. University of Cincinnati 2003.
150. V.V. Guliants, J.B. Benziger, S. Sundaresan, *J. Catal.* 1995, **156, 2,** 298.
151. V.V. Guliants, J.B. Benziger, S. Sundaresan, I.E. Wachs, J.-M. Jehng, *Chem. Mater.* 1995, **7, 8**, 1493.
152. V.V. Guliants, J.B. Benziger, S. Sundaresan, *Chem. Mater.* 1995, **7, 8**, 1485.
153. J.B. Benziger, V.V. Guliants, S. Sundaresan, *Catal. Today* 1997, **33, 1-3**, 49.
154. G.J. Hutchings, J.K. Bartley, J.M. Webster, J.A. Lopez-Sanchez, D.J. Gilbert, C.J. Kiely, A.F. Carley, S.M. Howdle, S. Sajip, S. Caldarelli, C. Rhodes, J.C. Volta, M. Poliakoff, *J. Catal.* 2001, **197**, 232.
155. G.J. Hutchings, J.A. Lopez-Sanchez, J.K. Bartley, J.M. Webster, A. Burrows, C.J. Kiely, A.F. Carley, C. Rhodes, M. Havecker, A. Knop-Gericke, R.W. Mayer, R. Schlogl, J.C. Volta, M. Poliakoff, *J. Catal.* 2002, **208**, 197.
156. K. Aitlachgar, A. Tuel, M. Brun, J.M. Herrmann, J.M. Krafft, J.R. Martin, J.C. Volta, M. Abon, *J. Catal.* 1998, **177**, 224.
157. U.R. Pillai, E. Sahle-Demessie, R.S. Varma, *Appl. Catal., A* 2003, **252, 1**, 1.
158. X. Wang, L. Xu, X. Chen, W. Ji, Q. Yan, Y. Chen, *J. Mol. Catal. A* 2003, **206, 1-2**, 261.

2

Use of Solid Catalysts in Promoting Water Treatment and Remediation Technologies

BY GABRIELE CENTI AND SIGLINDA PERATHONER

1 Introduction

The sustainable use of water is beginning to be recognized as a critical issue worldwide. As a consequence there is an increasing attention to this topic also within the catalysis community, because the use of catalysts may improve various water treatments or remediation technologies. As an indication of the increasing interest in this area, the number of publications in the journal *Applied Catalysis B: Environmental* has increased substantially. In the year 1995 four articles on the topic of solid catalysts applied to water treatment technologies were published out of a total of 54 articles. In 2003 the number of articles on catalytic water treatments jumped to 71 out of a total of 221 articles. Together with this four fold increase in the number of articles (corresponding to the large increase of interest on environmental catalysis), there was evidenced an increase of the share from 7% to 32% of the topic of catalytic water treatments. This thus becomes one of the main sub-topics of the area of environmental catalysis. The largest fraction of the articles were dealing on photocatalytic processes, but significant interest was given also to the catalytic wet oxidation using O_2, O_3 or H_2O_2 as oxidants, and to the conversions of nitrate and halo-hydrocarbons which are present as contaminants in ground and underground water. A limited number of articles dealt with less conventional methods such as the catalytic oxidation in supercritical water, sono- and sonophoto- catalytic methods, and electron beam and plasma catalytic methods, which, however, find major relevance in other journals out from the typical area of interest of the catalysis community. In general, a significant fraction (30-40%) of the articles dealing with the use of catalysts to promote waste water treatments were published not in journals specialing in catalysis, but more are now being published in catalysis journals.

1.1 General Aspects of Use of Solid Catalysts in Water Purification Technologies. – *1.1.1 Photocatalytic Treatments of Wastewater*. There are several excellent reviews on photocatalytic treatments of wastewater which include not only the use of TiO_2 as a photocatalyst, but also its combination with H_2O_2 or Fe^{3+}/H_2O_2 (the so-called photo-Fenton process) in order to improve the effi-

Catalysis, Volume 18

ciency of the process[1-5]. Photocatalytic methods can be considered a well established technology, although there is still need for improvement especially in terms of solar light response. Examples of application ranges from inorganic to organic compounds (chlorophenols, chlorinated solvents, pesticides and cyanide, etc.). The main limit regards the reaction rates and therefore the application is to mainly diluted waste solutions, although the solar photocatalytic treatment of municipal wastewaters (small communities) was also proposed[3]. Photocatalytic methods are certainly a very important technology, not only for water purification, but also for air purification (for example, for sterilization of air in hospitals).

Due to the variety of situations encountered in wastewater treatments (in terms of composition and concentration of waste, and effluent flow rates) there is no one preferable technology, but each may be suited in a certain range of concentrations and feed composition. Figure 1 reports a simplified technological map of applications of the different technologies for wastewater treatment, although it must be kept in mind that this map gives only rough indications and specific techno-economical evaluations have to be made.

There are some review articles which have compared the different type of AOPs [*Advanced Oxidation Processes*, which include heterogeneous TiO_2 photocatalysis, processes based on hydrogen peroxide (H_2O_2 +UV, Fenton, photo-Fenton and Fenton-like processes), photolysis, processes based on ozone (O_3, O_3 + UV and O_3 + catalyst) and on O_2 (wet air catalytic or non catalytic oxidation)] for defined high toxic and low biodegradable chemicals: pesticides[6] and chlorophenols[7]. These reviews are a valid source of references and their general conclusions are similar. Photocatalytic processes show significantly higher half reaction times than the rest of the AOPs for the treatment of most of the biodegradable chemicals, although there are also some clear advantages with respect to other AOPs. Similar conclusions are also given in the paper by Andreozzi et al.[8] which reports in addition some economic consideration about the parameters affecting the AOP costs. They conclude that AOP systems are not more expensive than the established technologies for pollutant abatement, but this conclusion is valid for limited fields of application were high purification levels are necessary. It should be remarked that, due to the growing knowledge on the area (especially on the use of catalysts in AOPs), the comparisons among these technologies is changing rapidly. In addition, water treatment and remediation comprises a large range of situations so that virtually each technology finds its area of application alone or in combination with other methods.

The comparison of the AOPs technologies is often made only in terms of efficiency in the removal of pollutants, but a more correct analysis should takes into account several parameters: (i) safety of operations, (ii) manageability and applicability to the specific operations (for example, when small volume effluents dispersed over a wide territory, it may be preferable to transport the water treatment unit instead rather than to collect all wastewater in a single central treatment unit), (iii) technical effectiveness using real streams, (iv) absence of necessity of post-treatments (to eliminate some residual compounds, even in traces, or metal ions), (v) secondary pollution prevention, (vi) minimal toxicity of

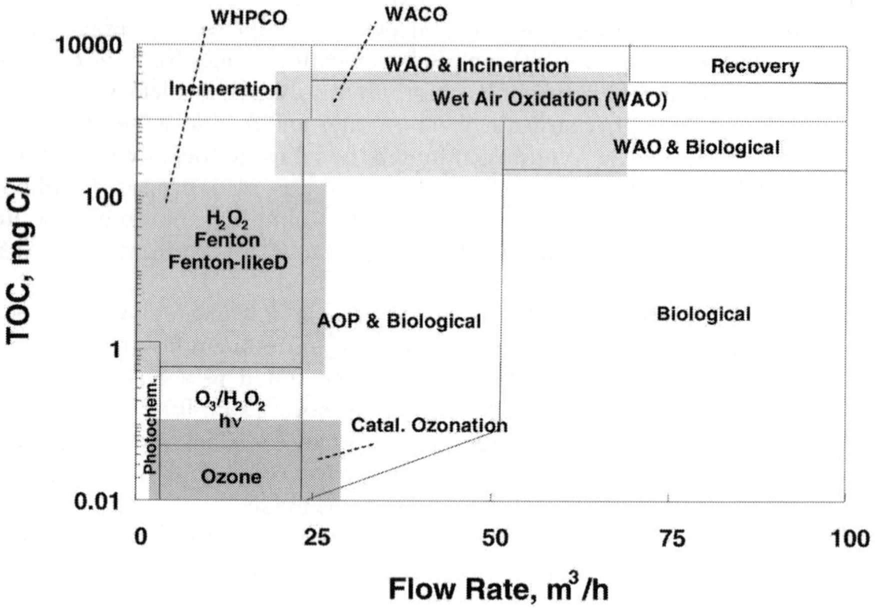

Figure 1 *Technological map for the applicability of different technologies of wastewater treatment. Elaborated from Centi and Perathoner*[64]

emissions, (vii) minimal corrosion, plugging, sensitive process-control parameters, and other operating difficulties, and not least (viii) cost. These parameters are often not considered or only in part.

There are several recent very interesting new areas in the photocatalytic removal of pollutants from water. In particular, two new direction of research may be cited, while for fundamental aspects regarding photocatalytic processes reference is given to the cited reviews[1-5]:

- The development of visible-light active photocatalysts,[9-11] a new exciting direction. With TiO_2 the light energy must exceed the band gap of 3.2 eV. Therefore, UV light where $\lambda < 387.5$ nm is required. By doping with transition metal ions or creating a defective TiO_2 structure researchers are searching how to improve the photocatalytic response using the visible light. A variation of this approach is the self-sensitivization of TiO_2 visible response by the reactants itself, when coloured chemicals are present in solution (this is the typical case of the elimination of dyes from textile wastewater)[12,13]. However, it should be noted that decolorization (conversion of the dye to non-coloured intermediates) is significantly faster than mineralization (conversion of the dye to CO_2), and when colour disappears the sensitizing effect of the dye is not longer present.
- The development of improved nano-structured and nano-composite photocatalysts. This is a new active area of research, because the microstructure of

titania has a marked influence on the photoefficiency[14]. Also nano-composites of titania with other materials such as zeolite[15], WO_3/SiC[16], β-Silicon carbide[17], carbon nanotubes[18], and carbon[19,20] were reported to allow significant enhancements in the photoreactivity.

A more detailed discussion of these aspects and in general of the use of catalysts in water treatment is reported elsewhere[21].

1.1.2 Wet Air Catalytic Oxidation (WACO). Wet air oxidation, e.g. the oxidation of organic compounds by oxygen at high temperatures and under pressure in autoclave reactors, is also a widely used technology, especially in the treatment of industrial wastewater or sludges. Use of catalysts has been extensively demonstrated to promote the reaction performance[22-25]. Combination of wet air catalytic oxidation (WACO) with a biological treatment has also been emphatized[26] and this appear as one of the preferable solutions in several examples of industrial waste water and sludge treatments[27]. It should be noted, however, that the number of industrial applications is still limited, both due to the time usually necessary for establishing new technologies, and for the critical issue of catalyst stability[28]. Wet air oxidation process (WAO), notwithstanding the name, operate typically under oxygen pressure (5-200 bar) (using air about four time higher pressure is necessary) and at elevated temperatures (125-320 °C). Residence times range from 15 to 120 min, but often longer times are necessary for high organics loadings or high levels of chemical oxygen demand (COD) removal (typically ranging between 75 and 90%). In contrast to supercritical water oxidation (SCWO), a complete mineralization of the waste stream is often impossible by WAO, since some low molecular weight oxygenated compounds (especially acetic and propionic acids, methanol, ethanol, and acetaldehyde) are resistant to oxidation. For instance, removal of acetic acid is usually minimal at temperatures lower than 300 °C. Organic nitrogen compounds are easily transformed into ammonia, which is also very stable in WAO conditions. In addition, there are gaseous emissions which should be also treated. Therefore, WAO is a pre-treatment of liquid wastes which requires additional treatments of the process liquid and gas streams. Over hundred plants are in operation today, mostly to treat waste streams from petrochemical, chemical and pharmaceutical industries as well as residual sludge from wastewater treatment[24]. The use of catalysts (WACO) allows to use milder reaction conditions[25], but especially to promote conversion of the reaction intermediates (for example, acetic acid[29] and ammonia[30]) which are very difficult to convert in the absence of catalysts, as mentioned above.

Figure 2 reports a simplified flow diagram of a WACO process which consists mainly of a high-pressure pump, an air or oxygen compressor, a heat-exchanger, a high-pressure (fixed bed) reactor and a downstream separator. The simplest reactor design is usually a cocurrent vertical bubble column with a height-to-diameter ratio in the range of 5-20[31]. A catalytic unit for the treatment of the off-gas is also typically necessary (Figure 2).

There are two main drawbacks of the WACO technology, besides the general

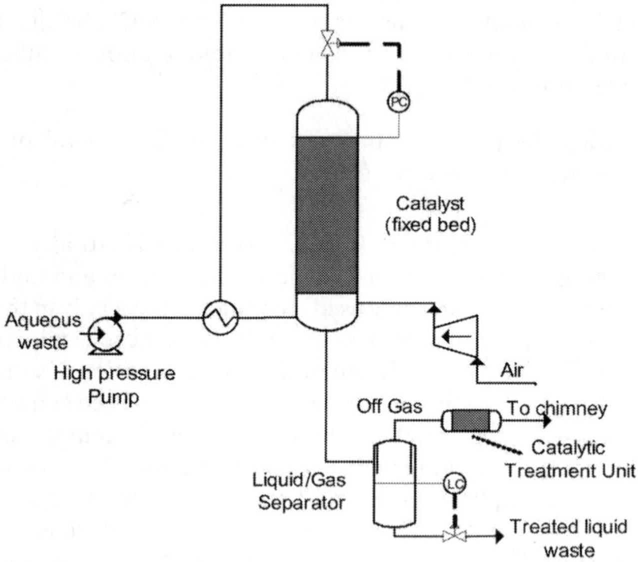

Figure 2 *Simplified flow diagrams of wet air catalytic oxidation (WACO) process*

issue of leaching of metals and catalyst deactivation common to all multiphase processes using solid catalysts:

(i) the cost due to high temperatures/pressures which makes energy consuming the process (the oxidation of organics is an exothermic reaction, but only in part the heat of reaction can be recovered, especially if a batch type autoclave reactor is used);

(ii) the cost of the reactor (it is necessary to use special materials such as titanium for the autoclave reactor, because there are often severe problems of corrosion related to the formation of high concentration of low molecular weight acids as the products of reaction; in addition, often the real streams to be treated contain large amounts of ions such as chlorine which may be oxidized under the conditions of wet oxidation, making very corrosive the reaction medium).

In addition, the use of high pressure reactors makes necessary the employment of specialized personnel and the use of all the typical safety procedures for a high pressure autoclave apparatus which further increase the costs. Although possible, the technology can be moved with difficulty from place to place due to safety related aspects. Additional issues are related to the treatment of discharged gaseous emissions containing NO_x, CO, VOC and smelling odours.

1.1.3 Catalytic Ozonation. Ozonation is a well established method especially for drinking water and is substituting in many applications the use of chlorine-based reactants for water sterilization. The interest in combining ozone with the use of solid catalysts has been recently increased and reviewed[32,32]. Catalytic ozonation was found to be effective for the removal of several organic compounds from drinking water and wastewater. However, catalytic ozonation has been studied mainly on a laboratory scale and there are still very limited commercial applications. One of the main problems in drinking water is the necessity of having virtually no contamination of the water from many metals which may act as catalytic elements.

Ozone decomposition in aqueous solution develops through the formation of OH radicals. In the reaction mechanism OH• radical has the role of initiator. The addition of hydrogen peroxide to the ozone aqueous solution enhances the O_3 decomposition with formation of OH radicals which are the effective oxidizing agents also using H_2O_2. The rate constant of reaction of OH radicals is in the order of 10^6-$10^9 M^{-1} s^{-1}$, several orders of magnitude higher than that of ozone or hydrogen peroxide. The pH of the solution has a great influence, since in the ozone decomposition mechanism the active species is the conjugate base HO_2^- whose concentration is strictly dependent upon pH. The presence of solid catalysts may alter local pH, but this question was scarcely considered up to now. However, this aspect can be a critical factor to understand and select catalysts to promote ozonation. There is a need for a better understanding of the effect of solid catalysts in modifying this radical-type aqueous chemistry and use these knowledge to develop tailored catalysts for ozonation.

1.1.4 Wet Hydrogen Peroxide Catalytic Oxidation (WHPCO). The wet hydrogen peroxide catalytic oxidation (WHPCO), often called also with a variety of names including Fenton oxidation, catalytic wet peroxide oxidation (CWPO), wet peroxidation, wet oxidation by hydrogen peroxide, etc., has been also discussed in detail in some reviews[18-39], but only in relation to the use of homogeneous catalysts. H_2O_2 is a powerful reactant, but needs to be activated to generate the OH• radicals which are the effective oxidizing agents. Activation requires either UV radiation or a catalyst. Fenton was the first who recognized more than a century ago[40] the effect of iron in promoting the reaction. This is the reason why oxidation processes utilizing activation of H_2O_2 by Fe (or Cu) salts are referred to as Fenton's oxidation. The photo-Fenton is instead a process in which Fe^{3+} salts and H_2O_2 are added to promote the performances of heterogeneous photocatalysts (TiO_2) or iron salts added to promote performances of H_2O_2/UV (UV-light water peroxide activation) system. The Fenton and photo-Fenton processes, notwithstanding the analogies, are distinct processes having different reaction mechanisms and influence of the reaction conditions. Fenton reagent (Fe^{3+}/H_2O_2) is used not only to treat waste water, but also to improve sludge dewaterability.

In a techno-economic comparison between different technologies to treat different industrial waste water streams[41] it was shown that the Fenton process is the preferable for treating refractory biodegradable industrial waste water streams. Balcioglu et al.[42] comparing photochemical and non-photochemical AOP methods for treating textile, Kraft bleaching, photoprocessing, and pharmaceutical wastewater showed the enhanced biodegradability using the Fenton method, although they also indicated that, depending on wastewater composition and applied pretreatment, all applied AOP increased the biodegradability of wastewater. The 'Environmental Technologies Action Plan Discussion Paper' of the European Commission (Aug. 2003) indicates the Fenton oxidation between the emerging fields of application for (a) domestic waste water treatment, (b) waste water reclamation and re-use (destruction and removal of harmful substances from the water streams to promote substantially the re-use of safe treated water in the industry or in agriculture), (c) sludge production reduction and (d) water disinfection. Therefore, the Fenton process appears rather interesting for a number of industrially relevant waste water streams, also taking into consideration the decreased cost of H_2O_2 as a reactant and its increasing use in waste water technologies.[43]

However, there are several drawbacks by using homogeneous Fenton catalysts: (i) possibility to work in a limited pH range around 3.5, (ii) difficulty in recovery or eliminating iron after reaction and especially (iii) the complexation of iron ions by reaction products such as oxalic acid which leads to a progressive lowering of the reaction rate and decrease of efficiency in H_2O_2 use as a function of the conversion. For this reason, there is an increasing interest in recent years in developing heterogeneous Fenton-like catalysts which can overcome these limitations[44]. A recent review[45] analyzes in details the recent progresses in this field.

The techno-economical comparison of WACO and WHPCO technologies suggests that the latter would preferable for the following situations, where capital costs are predominant over running costs:

- pre-treatment of the streams before to send to biological units which require the selective conversion of hardly biodegradable or toxic chemicals, but low levels of TOC (Total Organic Carbon) removal – less than 30% (for example, in the pretreatment of streams from agro-food industry),
- treatment of waste water having a medium-low content in organics (1-30 gC/l of TOC), and low volumes of effluents (1-100 m^3/day).

1.1.5 Selective Reduction of Pollutants in Water. In addition to oxidation treatments, a second large area of interest in waste water treatment regards reduction treatments and in particular the selective catalytic reduction of nitrates/nitrites to N_2 and the selective hydrodehalogenation of halogen-containing toxic chemicals such as halocarbons and pesticides. These technologies, which may find application both in wastewater treatment and in water remediation technologies (see later), show distinct advantages over conventional methods, because available technologies are often inadequate to meet economic and ecological demands. In addition the commercial technologies often require large centralized treatment units. In many cases, wells serve small local communities which cannot be connected to centralized water treatment units. It is thus necessary to develop technologies which are compact, transportable and easily manageable. Catalysis plays a central role in developing this kind of application[46].

Nitrate and nitrite selective catalytic reduction has been discussed in various review articles[47-50]. Interest on this technology is increasing, although for drinking water applications the selectivity to ammonium ion formation (which concentration should be below 0.5 ppm) and the possible contamination even in traces by Pd (the active catalytic element) remains an issue. These problems, however, are less critical for other industrial applications.

Excess intake of nitrate ions could cause human problems, because they can be easily reduced to nitrites in the intestines causing serious health problems particularly for infants (blue baby syndrome). Furthermore, nitrites are the precursors to the carcinogenic nitrosamines as well as to other N-nitroso compounds. Present European Union limits indicate nitrate and nitrite concentrations of 50 mg/L and 0.1 mg/L, respectively, but new European standards guideline indicate a value of 25 ± 5 mg of nitrate per litre. Several techniques for eliminating nitrates in producing drinking water are available at present. The two most used methods are (i) physicochemical processes (ion exchange and membrane techniques), and (ii) biological processes (heterotrophic or autotrophic techniques, the latter being the most used).

Autotrophic denitrification occurs by the raw water percolating through a porous thionic material to which 'Thiobacillus denitrificans' type bacteria are attached. Autotrophic bacteria synthesize their compounds from mineral carbon in the water in the form of hydrogen carbonate (HCO_3^-). Heterotrophic bacteria synthesize their compounds from organic carbon (ethanol, acetic acid), generally exogenous in raw water. Heterotrophic denitrification occurs by the raw water percolating through a porous material to which is attached Acinetobacter, Pseudomanas, Alcaligenes, Spirillus bacteria which reduce the nitrates in gaseous nitrogen (N_2) in anaerobic conditions.

Resin denitration is the most widely used specific treatment for nitrate removal. It tolerates variations of temperature and flow rate and is rather easy to operate as it is more stable over time. However, the disadvantage of this technique is the generation of waste (brine) containing various anions. The system requires periodic regeneration, is very sensitive to suspended matter (resin clogging) and extremely sensitive to organic matter (irreversible deactivation). Furthermore, is not effective in the presence of large amounts of other anions (Cl^-, $SO_4^=$, HCO_3^-). Membrane (reverse osmosis or electrodialysis) allows a continuous technology, but also produces brine. The waste brines produced by reverse osmosis can amount up to 25% of feed water. Ion exchange and membrane technologies only concentrates nitrate, but not eliminate them. They are thus less acceptable on an environment point of view, and applicable when the released solution concentrated in nitrate ions cannot cause environmental problems (for example, when discharged in sea water). Membrane disadvantages are also noise production and high energy requirements.

Biological denitrification is more complex to operate because it relies on the capabilities of active bacteria, but effectively convert nitrate to not toxic chemicals (N_2). It can be subject to transient states generating malfunctions and requires thus greater care. Biological denitrification produces bacteria sludge, requires chemical additives (carbon source and other nutrients for bacteria) and especially requires large installations.

The alternative possibility is to use a catalytic reduction of nitrate[47-50]. Nitrate ions are reduced on a supported Pd-Me catalysts, where Me is Cu, Sn or other transition metals, by using H_2 as the reductant. The technology allows a virtually waste-free process that can be scaled from small to large application; there is no noise production and has relatively low energy requirements.

Table 1 reports an overview of the advantages and comparative characteristics of these technologies for water remediation by nitrate contamination. An estimation of the costs for the technologies is also given, but must be considered that the cost may considerably depend on the amount of water to treat, the composition of water, etc. Furthermore, in the case of catalytic reduction no commercially available technology exists. The sensitivity to down-sizing is also given, because as cited before there is a need especially of technologies to treat relatively small volumes of water.

Considering a technology which should be applied to water remediation by nitrate contamination and to be used in a small community agricultural area, the catalytic reduction offers better balanced advantages in terms of eco-compatibility, manageability and multiple use (for example, can be moved to be used for different wells), although cost may be slightly higher. Furthermore, the same technology can be applied to the simultaneous removal of pesticides and halogenated compounds (see below), while the other technologies are not effective. When water remediation requires the contemporaneous removal of nitrate and halogenated compounds, and technology should be used to treat small water sources (less than 100-200 m^3/day), the catalytic technology appears a preferable solution.

Hydrodehalogenation is another area of increasing interest in water treatment and remediation. It is interesting to note that catalysts active for this reaction show several analogies with those active in the catalytic denitrification discussed above. Often, underground water is contaminated by high concentration of nitrate contains and halocompounds (pesticides, for example) in concentration above the allowed regulations. The activity of palladium-based catalysts for treating groundwater contaminated with halogenated hydrocarbons by hydrodechlorination with dissolved hydrogen has been reported[51-54]. It was also already shown that notwithstanding the different reaction rates, the same catalysts may be used for the combined removal of nitrate and pesticides from contaminated water[54]. However, results reported in literature are limited. This is an area which should be studied in a more in depth in the future.

1.1.6 Catalytic Permeable Reactive Barriers (CPRBs). There are other areas related to water treatment and remediation in which the use of solid catalysts improves performance, although these areas are less investigated than those discussed earlier. One example is the use of solid catalysts in catalytic permeable reactive barriers (CPRBs). Permeable reactive barriers are emerging as a very interesting alternative for both organic and inorganic water contamination[55]. The main advantage of a reactive barrier is the passive nature of the treatment and therefore the low cost. Once installed, the barrier takes advantage of the in situ groundwater flow to bring the contaminants in contact with the reactive material. Solid catalysts have been shown in some cases to significantly promote

Table 1 *Comparison of the characteristics of technologies for water remediation by nitrate contamination. Elaborated from Centi and Perathoner*[46]

Technology	*Ion Exchange*	*Reverse Osmosis*	*Biological Denitrification*	*Catalytic Reduction*
Indicative Cost* (€/m^3)	0.15-0.25	0.4-0.6	0.2-0.3	0.25-0.55
Sensitivity of costs to scale-down	Medium	High	High	Low
% Efficiency in water purification	85-98%	75-80%	98%	98-100%
Multipurpouse use†	None	Depending on molecular size	Some‡	Highly effective
Fate of nitrate	Adsorbed & concentrated	Concentrated in a waste stream	Transformed to N_2	Transformed to N_2
Waste	Waste Brine	Waste Brine	Bacteria Sludge	None
Chemical Additives	Sodium Chloride	Sulphuric acid and base	Ethanol and phosphoric acid	H_2
Flexibility in variable operations	Medium	Medium	Low	High
Energy use	Low	High	Medium	Low
Space requirements	Limited	Limited	High	Low
Movable	Yes	Yes	No	Yes
Manageability	Good	Good	Low	Good
Type of operations	Periodic regeneration	Continuous	Continuous (but not during transient)	Continuous
Sensitivity to deactivation	Medium	High	High	Medium
Automatic control	Simple	Simple	Complex	Simple
Start up time	Immediate	Immediate	Up to 1 month	Immediate
Monitoring required	Little	Little	Intensive	Medium
Selectivity of the process	Low	Low	High	High
Odours	No	No	Yes	No
Noise	Some	High	None	None

* Estimated costs for a 1000 m^3/d process. † Removal of pesticides and other halogenated compounds. ‡ Sensitivity to pesticides.

the performances of PRBs. Fry et al.[56] reported the development of catalysts (supported metal porphyrins) for dehalogenation of some EPA contaminant list compounds in PRBs. The use of zeolites to promote the performances of PRBs in MTBE conversion has been also reported[46,57].

Methyl *tert*-butyl ether (MTBE) is a widely used gasoline additive, but due to its contamination of both ground and surface water, its use in US has been banned recently. MTBE water contamination derives from leaking of underground fuel tanks and pipelines, tank overfilling and faulty construction at gas stations. While other gasoline components are characterized by good biodegradability and low water solubility (thus gasoline release shows a fast 'natural attenuation' and low migration index), MTBE is quite soluble in water (therefore, highly mobile in ground water systems) and resistant to biodegradation. Furthermore, a long incubation time is usually necessary before the start of microbiological degradation. Acid zeolites with suitable pore structures (channel structure and pore openings) such as H-ZSM-5 and H-BEA can be used as catalytic permeable reactive barriers for in situ remediation. These zeolites not only act as adsorbents for both MTBE and its reaction products, but also effectively catalyze the hydrolysis of MTBE to t-butyl alcohol (TBA) and methanol (MeOH) which then can be rapidly biodegraded by indigenous microorganisms.

Figure 3 reports the effect of a H-ZSM-5 zeolite in catalyzing the hydrolysis of MTBE in the presence or absence of the stirring of the solution[57] and for comparison the hydrolysis in the absence of the zeolite and when the aerobic MTBE degradation by the indigenous micro-organisms present in the underground water starts. It is evident that the zeolite has a marked effect on the hydrolysis of MTBE which allows it also to improve the rate of bio-degradation, because the products of hydrolysis (methanol and t-buthyl alcohol) are biodegraded much faster than MTBE.

1.1.7 Use of Heterogeneous Catalysts in Other Waste Water Treatment Technologies. Concluding this short overview on the use of solid catalysts in promoting water treatment and remediation technologies, less conventional applications should be also briefly mentioned. Supercritical water oxidation (SCWO), e.g. at reaction conditions that exceed the critical point of water (374°C, 218 atm), is receiving an increasing interest for the possibility of effective destruction of organic wastes in a wide variety of potential applications including waste treatment and resource recovery for manned space exploration, destruction of high-risk military wastes, treatment of pulp and paper mill and municipal sludges, and treatment of wastes in a commercial chemical facility[58]. The chief processing advantage of SCWO relative to wet oxidation at milder conditions is that the oxidation reaction can go fast to completion (no secondary treatment is needed), but the cost of high pressure/temperatures and the critical aspects related to corrosion of the autoclave materials are issue swhich limit the application of the technology. The economics of catalytic SCWO has been assessed and compared with conventional SCWO, incineration, and wet oxidation[59]. The analysis revealed that a range of waste stream compositions exists for which catalytic

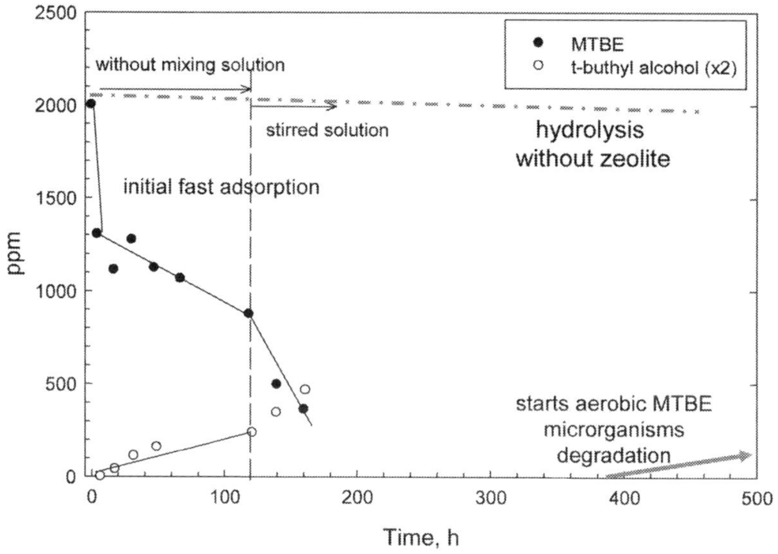

Figure 3 *Behavior of H-ZSM-5 zeolite (SiO_2/Al_2O_3 = 25) in the room temperature conversion of MTBE in the presence and absence of mixing of the aqueous solution containing an initial concentration of 2000 ppm MTBE. For comparison the hydrolysis in the absence of the zeolite and when starts the aerobic MTBE degradation by indigenous micro-organisms present in underground water is also reported. Elaborated from Centi el al.*[57]

SCWO is likely to be the preferable method.

The largest part of the studies reported in the literature deals with homogeneous fluid-phase SCWO. However, there is an increasing interest recently in using solid catalysts to further improve reaction rates and lower costs. This catalytic SCWO technology would combine the advantages of catalytic oxidation (faster reaction rates) and supercritical reaction conditions (a single fluid phase)[60]. Heterogeneous solid catalysts are preferred to avoid a catalyst-fluid separation step. Yu and Savage[60b] recently discussed the use of bulk MnO_2, bulk TiO_2, and CuO/Al_2O_3 catalysts for the oxidation of phenol in supercritical water. MnO_2 gives the highest oxidation rate on surface area basis (second-highest mass-based rate), and maintains its activity with use, and is stable under SCWO conditions, different from CuO/Al_2O_3 catalysts for example.

Although the cited article reports some interesting results and opens some possibilities of application, there is few attempts to tailor catalysts for this specific application. This is a common limit often found in several articles dealing on the use of catalysts for water treatment, due to the lack of fundamental knowledge on the chemistry of these catalytic reactions.

A second remark, also common with many articles on catalytic treatments of waste water, is the use of phenol as model molecule. Phenol conversion is quite easy and often do not provides reliable data even for screening purpose. Therefore, an effort should be made in going from very simple to more complex model molecules and possibly also to realistic waste water streams.

Another interesting article reporting the use of heterogeneous catalysts in SCWO was published by Martin et al.[61] who investigated the conversion of ethyl acetate (as model compound for polyesters) using a heterogeneous MnO_2-CuO/Al_2O_3 catalyst. Without a catalyst this compound can be hydrolyzed easy in sub-critical conditions, but the catalyst promotes the rate of oxidation, which in super-critical conditions becomes ten fold higher than in sub-critical conditions. The application of a catalyst leads to a significant change in the reaction pathways favouring deep oxidation. They also observed high stability of the catalyst, but also the loss of Cr from the stainless steel reactor walls. Reactor corrosion is claimed to be less severe in SCWO conditions than in WACO conditions, although not fully proven.

Sonochemical degradation of organic pollutants is another interesting not-conventional technology for water treatment. The chemical effects of high-intensity ultrasound result primarily from acoustic cavitation: the formation, growth, and implosive collapse of bubbles in liquids. Acoustic cavitation generates sites of locally high temperatures and pressures for short period of time, which give rise to the H_2O sonolysis with the production of radical species (H•, HO•, HOO•) and direct destruction of solute[62,63]. Some attempts has been direct to study the effect of heterogeneous catalysts during the sonochemical degradation of organic pollutants recently. Ge and Qu[64] have investigated the degradation of azo dye Acid Red B on MnO_2 enhanced by ultrasonic irradiation. Ultrasonic irradiation enhances the degradation of the dye both for decolorization and mineralization degradation via both physical and chemical effects of sonication. Probably, ultrasonic irradiation induces the removal or deformation of the passivating outer oxide layer normally present on MnO_2 surfaces via its microjetting effects or shockwave damages. Shockwave and microjetting produce lattice defects and surface vacancies which accelerate the MnO_2 reactivity. On the other hand, the creation of structural defects by ultrasound generates H_2O_2 on the catalyst surface which can enhance the reaction rate, but also the dissolution of MnO_2. One limit of this technique is therefore water contamination and the progressive deactivation of the catalyst. Ince et al.[65] have reviewed recently the removal of refractory organics in treated effluents or natural waters catalyzed by ultrasonic pressure waves, with emphasis on basic theories, physical/chemical principles, state of the art, and figures-of-merits of sonochemistry.

The use of ultrasound to enhance performances of titania (or preferably TiO_2-loaded zeolites and mesoporous materials) have been instead reviewed recently by Smirniotis et al.[66] It was found that the presence of an ultrasonic field enhances the rate of photodegradation. A threshold of specific energy that should be provided in the reaction medium exists. This threshold is about 0.1 kW/l and operation under this volume will ensure structural stability of these catalysts used. The use of ultrasound also avoids the formation of toxic intermediates observed when photocatalytic degradation of phenol alone is used.[67]

Plasmacatalytic methods (non-thermal plasma generated by low energy corona discharge or microwave coupled with solid catalysts) in an area of growing importance for auto exhaust emissions (low-temperature conversion of NO_x to

N_2 in excess oxygen, H_2 generation) and for the oxidative removal of volatile organic compounds (VOCs)[68-70]. There is some attempt to use these methods for waste water treatment. Wekhof and Gudernatsch reported[71] the use of non-thermal plasma (NTP) to treat industrial wastewater. The treatment surpasses by far the conventional UV wastewater treatment in eliminating micro-organisms, viruses, and organic compounds. Malik et al.[72a] reviewed recent developments for disinfecting water using electrical discharge processes, particularly pulsed corona discharge, dielectric barrier discharge, and contact glow discharge electrolysis techniques (pulsed corona, dielectric barrier discharge, and contact glow discharge). The same author[72b] reported also about the use of α-alumina, γ-alumina, and silica gel in a pulsed corona discharge (PCD) reactor to increase the rate of decomposition of aqueous methylene blue. Methylene blue concentration in the effluent was reduced to 23% of the inlet concentration by PCDs in water. Under the same experimental conditions, addition of α-alumina further reduced it to 8%, γ-alumina to 4%, and silica gel to below the detection limits. PCDs with silica gel were run for > 100h in a continuous flow reactor and methylene blue in the effluent remained below the detection limit. Shimizu et al.[73] also pointed out the potential use of microwave-induced plasma for treating waste water and the effect of solid catalyst to trigger the nature of the radical species observed in the generated plasma. SiC and carbon block favour O and OH radicals, while perovskites such as $Lao_{0.8}Sr_{0.2}CoO_3$ or $La_{0.6}Sr_{0.4}MnO_3$ favour the preferential formation of H radicals.

Water treatment using electron beams is another area of growing interest[74]. The effectiveness of e-beam treatment results from its production of huge concentrations of highly reactive radicals ($OH^\bullet$, $H^\bullet$) and aqueous electrons (e_{aq}^-) through the ionization of the high energy electrons as they cascade within the water. These short-living radicals drive both oxidation and reduction reactions at the same time. Since the digestion of organic hydrocarbons typically requires a sequence of reactions of both types, e-beam treatment uniquely has the capability to drive digestion through several intermediaries to non-toxic end products. A new generation of pulsed linear induction accelerators for applications in water treatment has been introduced recently. These accelerators are compact, modular, and produce high average power that enables them to treat high-flow rate streams. This process has a number of applications, including industrial water pollutants, drinking water, municipal water, and sterilization and disinfection. Although specific studies on the use of solid catalysts together with e-beam irradiation are still missing, it is reasonable that similar to what discussed for NTP, several possible advantages in combining solid catalysts with these sources of free radicals and/or electrons could be obtained. This will allow to reduce energy costs (one of the main drawbacks of these technologies, together with safety aspects and the high fixed costs) and optimize the performances.

The use of catalysts together with NTP or e-beams can be thus a promising technique, although the results are still at an initial stage. Using solid catalysts the chemistry of generation of radical/charged species can be significantly influenced and tuned, in addition to the role of the solid catalyst in enhancing the rate of reaction by adsorbing the pollutants (a relevant engineering aspect also, because this would allow to have irradiation on a defined and limited area). However, it is also evident that new concepts should be explored in relation to the use of solid catalysts with highly reactive radicals. These new directions of research for wastewater treatments also open new frontiers for understanding and applying solid catalysts which may result in new breakthrough applications also out of this specific field of application for the development of novel low temperature catalytic processes, a necessary direction toward novel processes for a sustainable (green) chemistry[75].

1.2 Perspectives in Using Solid Catalysts for the Treatment of Water. – The use of catalysis for wastewater treatments may be considered still at the initial stage, even though there are a few examples of industrial applications. The most relevant is the 'Osaka Gas' wet catalytic oxidation process for the treatment of high-concentration industrial effluents[24]. The process may be applied to wastewater from the chemical and pharmaceutical industries, pulp and food production, and gasification processes, the latter application being the original scope (coal gasifier effluents). The process operates in the 200-250°C range using precious metals supported on titania or titania-zirconia washcoat over ceramic monoliths and is tailored to give a high conversion of ammonia and cyanide (present in the coal gasifier effluents). This is actually the only industrial process of wastewater treatment which applies a solid catalyst, while there are some examples of industrial processes of wet oxidation where transition metal ions (iron and copper, in particular) are added to promote activity. The use of catalysts in AOP processes offers several potential advantages :

- *Operation in milder reaction conditions.* Besides being less expensive and allowing relevant energy savings, carrying out wet oxidation at lower temperatures and pressures is of critical importance to reduce the possibility of corrosion associated with the formation of acids in the process of mineralization of the organics. This allows less costly materials to be used for the autoclave as well as safer operation.
- *Better engineering.* More compact reactors, elimination of unwanted byproducts even if in traces, better reactor control and heat recovery can be achieved by advances in engineering.
- *Possibility of selective conversion.* Using catalysts it is possible to achieve the preferential conversion of some chemicals, although this concept has not yet really been analysed in detail. However, it is of critical relevance. For example, when an AOP method is used to pretreat wastewater before to be send to a biological unit, the aim is to selectively convert low biodegradable or toxic chemicals, but not other compounds, because this would increase costs.
- *Possibility of combination of removal by adsorption and catalytic conversion in a separate step.* Solid catalysts may act as adsorbent to remove pollutants present in diluted wastewater emissions, and act as catalysts to catalyse the conversion of adsorbed substances during a regeneration step[76]. This allows much better process energy use than wet (catalytic) oxidation for diluted wastewater. Also active carbon can be functionalized with a catalyst[77] to allow its regeneration by oxidative treatment at much lower temperatures (around

250°C) than those used in the conventional regeneration method for active carbons (around 800°C)

The use of catalysts, furthermore, is required in the processes of wastewater purification by reducing treatments. Catalysts also find also application as complementary technologies to other wastewater treatment methods, such as in the control of odour, VOC, N_2O and NO_x emissions from wet oxidation treatments (for example, in the wet oxidation of industrial sludges), and of odours and VOC emitted from biological processes (aerobic and anaerobic). Although usually commercial catalysts are used in these cases, there are often unpredicted effects in treating complex mixtures and thus more specific catalysts would be preferable. The same is valid for catalysts used to convert stripped VOC from contaminated groundwater.

The use of solid catalysts to improve technologies for wastewater treatment is an area of growing relevance, but further fundamental knowledge is required, especially on the following aspects:

- stability of the catalyst under conditions of wastewater treatment (especially in relation to leaching of transition metal active components)
- synthesis of catalysts not containing toxic elements the leaching of which, even if in traces, may prevent the applicability of the technology according to current legislation
- optimisation of the catalyst texture and pore structure, because of the mass transfer problems which are completely different from the case of gas-phase reactions
- relationship between surface catalyst properties (acido-base properties and their modification by adsorption of reactants, hydro-phobic/phylic properties, formation of double electric layers), and rate of diffusion of radical and charged species from the bulk to catalyst surface (and inside pores)
- reactor engineering, new advanced reactor engineering solutions are required in catalytic wastewater treatments.

In the largest part of the studies, few attempts have been made to tailor solid catalysts for the water treatment applications. There are problems of pore structure design (intraparticle diffusivity is usually an issue determining the catalytic performances), leaching (often driven from the reaction products and thus depending on the progress of the reaction), stability (related, but not only depending on the possible leaching), and selectivity (typically, quite complex mixtures should be treated in practical applications). As a consequence catalyst design for water treatment is different from that of most of other applications and even in the screening of the sample this aspect should be considered, as perhaps in the development of all multiphase reaction processes[78]. New micro- and nano-structured materials (fibers, nanotubes, multilayer composite materials) are available for these reactions.[78] This requires a better understanding of these aspects and therefore opens new perspectives of investigation.

There is an even more challenging area. Because it is often necessary to eliminate pollutants from large volume of water, it is necessary to avoid heating of these streams to minimize energy costs. This is the reason why either highly reactive chemicals (ozone, hydrogen peroxide) or external energy sources (UV light, microwaves, ultrasounds, etc.) are used to generate the radical species (H•, HO•, HO_2•, etc.) which are the effective oxidizing species. There is little knowledge on the relationship between solid catalyst properties and reactivity in generating or scavenging these radical species, apart for the case of titania photocatalyst. It is thus evident that a better understanding of these aspects is necessary which may open new directions for research also for other fields of using solid catalysts.

2 Water Issue and the Role of Solid Catalysis in Promoting New Technologies

2.1 Background. – Water is needed to generate and sustain wealth as well as for recreation. Water is often seen as an abundant and cheap resource, but instead large regions in Europe and in the world suffer of water stress and concerns[80]. This problem derives from a combination of different factors such as overexploitation of available water, changed and depleted natural vegetation, reduced water infiltration into the soil and increased surface water run-off, and finally pollution. The overexploitation of water resources often combines with a poor quality of water. In fact, the increasing contamination of water sources by human and industrial activities requires increasingly longer times for the natural attenuation of contamination, but in contrast low water resources imply a shorter time for water cycle. Furthermore, a depletion of natural vegetation causes a lowering of the phytoremediation capacity of soil as well as a reduced effectiveness of crops in capturing fertilizers and pesticides. In many regions there is thus a combined lowering of water resources and of water quality. It should be also considered that in many costal areas, overexploitation of water sources causes also salt water intrusion in aquifers which may remain contaminated for long periods. Water treatment technologies have a long history of more than a century, but the combined action of increased use of water, lowering of the amount of available fresh water and increasing quality requirements (legislation in every country imposes environmental regulations and health quality standards that steadily become more restrictive) has renewed the need new and/or improved technologies in recent years[80,81]. In fact, water and wastewater has become a major social, technological, economical, and political problem. As an example, Figure 4 reports a distribution map of nitrate in groundwater and PCB (polychlorinated biphenyls) in Europe. Over 30% of the available groundwater for human uses results to be contaminated with levels of these pollutants above the suggested limits in the concentration. Moreover, disposal of sludges poses additional problems.

Water consumption in industry has also increased considerably in the last decade and at the same time the cost of industrial water, which includes its direct cost, the cost of the treatments necessary to reach the water quality requested for the specific application, and/or the cost of post-treatments, has increased ex-

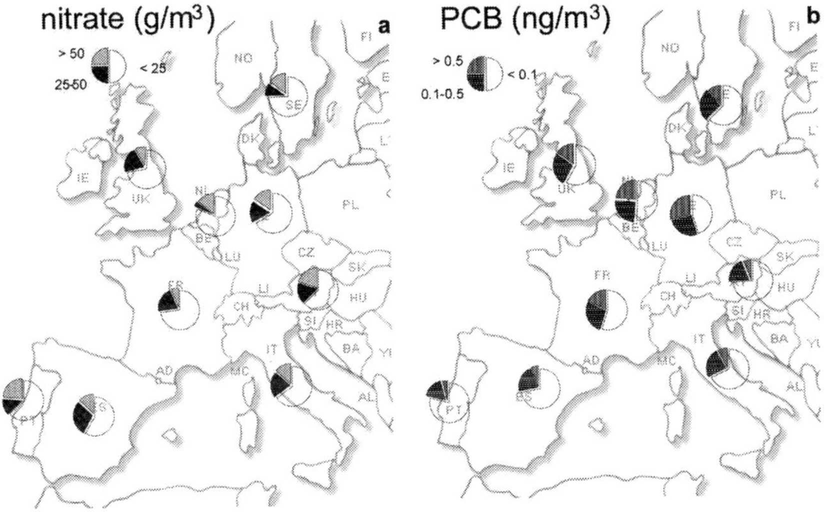

Figure 4 *Distribution map of nitrate in groundwater and PCB (polychlorinated biphenyls) in Europe. EEA (http://www.eea.eu.int/) and EMEP (http://www.emep.int/) sources. Elaborated from Centi and Perathoner*[64]

ponentially. In several industries large amounts of water are used in cleaning and process applications and thus recycling this water can be an opportunity to combine a reduction in the costs of industrial water with improved control of water management and a better environmental impact on natural resources[27,82].

There is thus an increasing need of technologies to restore the required water quality. Different uses such as for drinking water, recreation, industrial use and for agricultural uses such as irrigation and livestock watering require different water quality. The choice of water treatment technologies depends thus on both the characteristics of water source and the quality necessary for final use[81]. However, several additional factors determine the choice of water treatment technologies: cost, manageability, impact on environment, possibility of automatic control, social acceptance.

2.2 Water Treatment Technologies. – The water treatment technologies are differentiated between water treatment technologies and water remediation technologies. Although there are several analogies between the two cases, and often similar technologies may be used, the differentiation is related to the different characteristics of the water to be treated which are in relation to the different characteristics of pollution which may be broadly classified according to point and diffuse pollution sources. Point sources of pollution are specific and readily identifiable, such as discharges from sewage treatment works and industrial processes. Diffuse sources are less clearly defined, e.g. run-off from agricultural land and urban areas, and pollution from waste disposal, roads and railways, spills from tanks, etc. The term 'technologies for water treatment' refers to the purification of water streams deriving from point sources, while the term 'technologies for water remediation' to the purification of ground or underground water contaminated by diffuse pollution. In the former case, the technology has a series of constraints given by the specific application. For example, a water treatment technology for the purification of industrial wastewater should be characterized by Liquid Hourly Space Velocities compatible with the upstream process, or should be able to handle fluctuations in the composition of the pollutants (a common problem). In water remediation technologies such constraints often do not exist, but there are other constraints such as costs, manageability and transportability of the technology, etc. As a consequence, even for similar cases of removal of pollutants from water (for example, nitrate from underground water or from industrial wastewater) the selected technology for water treatment may be different.

A different case regards the technologies for drinking water. In fact, water quality for drinking purposes is often much more stringent in terms of both chemical and biological contamination than for other civil or industrial applications, but on the other hand other components that should be removed in the first case (for example, carbonate in water to be used for boilers) may be not removed.

2.3 Technological Needs for Sustainable Water. – Although progress has been made in the last decade, sustainable water use requires a change in the approach to water and wastewater management, but also advances in scientific and technological knowledge. In terms of scientific advances, critical topics are the understanding of the impact of existing pollutants (and of emerging ones), including improved modelling techniques and analytical tools, and of new water management approaches. But of equal relevance is also the need for developing new advanced technologies. In terms of water quality issues, there are problems which are progressively reduced with current methodologies/technologies such as those of phosphorus levels, acidification of surface water, and contamination by heavy metals and those which need further effort in developing better and sustainable technologies. These include, but are not limited to :

- reduction of the contamination of groundwater and especially drinking water by nitrates, and by pesticides (or other Persistant Organic Pollutants – POPs) and chlorinated hydrocarbons;
- improved technologies which allow reuse of (industrial) wastewater;
- advanced technologies for the elimination of chemicals having a toxic effect on microbiological activity from diffuse sources such as elimination of polyphenol emissions in agro-food production. Similar relevant issue is the elimination of antibiotics (largely used in animal feed) which concentration is increasing in ground water of agricultural areas and cause a significant alteration of biodiversity.

Sustainable water technologies, besides being economic and easily manageable should avoid secondary pollution or the simple transfer of the pollutants from one stream to another.

Water treatment technologies are often multistep processes which include:

1. *Primary treatments.* These are based mainly on physical action; examples are screening, sedimentation, filtration and flotation.
2. *Secondary treatments.* These are based on chemical and biological processes which convert the chemical structure of the constituents of wastewater to products either innocuous or less hazard; examples are neutralization, precipitation, and active sludge treatment.
3. *Tertiary treatments.*These are used to convert chemicals difficult to convert or not converted by secondary treatments, and when more stringent water quality is required; examples are active carbon adsorption, ion exchange, reverse osmosis and the AOPs.

The typical efficiency in pollutant removal is less than 50% for primary treatments, in the 70-90% range for secondary treatments and over 95% for tertiary treatments. While primary and secondary treatments are well established, the improvement of final water quality in cleanup technologies and therefore the reduction of its impact on the environment and global water cycle requires the use and further development of tertiary treatments. Typical tertiary treatments such as ion-exchange, reverse osmosis, active carbon adsorption, and electrodialysis do not convert pollutants, but rather transfer them from a diluted to a concentred stream. Therefore, it is preferable in a sustainable water approach to use other technologies such as AOP methods which convert pollutants and toxic chemicals to innocuous or not hazardous products, although these methods are usually more expensive and require further development. There is thus a technological need for further developing these technologies and not only for their more extensive application.

Furthermore, these methods find application in combination with other treatments in converting recalcitrant chemicals or in improving water quality in the treatment of a wide range of industrial wastewater effluents. For example, they may be used for converting toxic compounds to chemicals (such as acetic acid, for example) which then can be treated by active sludge in a biological downstream unit.

Although active sludge remains the main technology to reduce the BOD (Biological Oxygen Demand, an index of the content of biodegradable organic substances in water) in wastewater, new technologies are required which can be more flexible in terms of organic loading (often, the depuration plant is bypassed and the effluent sent directly to the river when there is peak in the organic content of the wastewater) and content of toxic elements (various chemicals such as phenols, cyanide, etc. and heavy metals are toxic for microrganisms). There is, however, a new driving force which is related to social acceptance of wastewater treatment plants. Typical active sludge plants have a visual negative impact and produce odours which often makes difficult (sometimes impossible) their localization in, or near cities. New advanced technologies may hide away the wastewater plant in an office-type block which makes its acceptance by society much easier.

Another problem related to active sludge treatments is the disposal of the sludge. For example, the EU Urban Wastewater Treatment Directive has put severe limitations on the landfill disposal, typical method up to now used (after stabilization, dewatering and thickening). Alternatives are incineration, which is a costly method and may produce relevant amounts of gaseous pollutants, or wet oxidation, either catalytic or not catalytic[84].

2.4 The Issue of Water Recycle in Industry. – There are several types of production which have an intensive use of water, such as automobile and heavy manufacturing, agro-foods and dairy products, the electronics industry and chemical and paper manufacturing. Water is used in processing, washing, cooling and as service water. Water contamination and therefore the possibility of its recycle greatly depends on the specific treatment. Water used in cooling and as service water may be easily recycled, while processing and washing water requires one or more treatments before recycle. Due to the large volumes of water used in washing or processing in some companies, an effort in this direction is mandatory. However, processing and washing water often is characterized by the presence of several toxic or non-biodegradable chemicals. Typical solutions adopted are ion-exchange, reverse osmosis, active carbon adsorption, and electrodialysis, but the development of more sustainable technologies is necessary.

Large efforts have been made by several companies to recycle water. Water recycle, besides reducing the use of water resources and allowing better water management in the company, can also be a relevant economic opportunity, especially when high water quality is required for the production[82]. As discussed in more detail in the cited reference[82], rinse water recycle in printed boards production, semiconductor fabrication and the textile dying industry can lower the cost of water by about 5-15% which, taking into account the large volumes of water used in these processes (typically from 100,000 m^3/y to up to 1 million m^3/y), may represent a very large economic opportunity, besides reducing the negative impact on water resources.

Typically rinse water in the electronics industry is produced in the several steps of (i) cleaning (removal of oil, particles and other contaminants), (ii) etching, (iii) layer deposition and (iv) layer removal. The composition of the rinse water is very complex, due to the presence of different chemicals, although in low concentrations, such as (1) organic acids (methane sulfonic acid, p-toluensulfonic acid, formic acid, citric acid, etc.), (2) organic bases (triethanolamine, thiourea, amines, etc.), (3) surfactants (polyglycol ether, alkyl benzene and alkanes sulfonates, α-sulfo fatty acid esters, soap, fatty alcohol polyglycol ether, etc.), (5) complexing, wetting agents and organic stripping agents (cyanides, tartrate, polyethylene oxide, etc.), (5) other organics (alcohols, resorcinol, γ-butirrolactone, formaldehyde, etc.), and (6) inorganics (H_2SO_4, HBF_4, HCl, K-citrate, NaF, NH_4Cl, Cu, Sn, Ni, Pb).

The development of technologies to recycle this rinse water requires an intense research effort and the combination of different water treatment technologies. In the specific case of recycling rinse water for the production of printed circuit boards (PCBs) and semiconductors[82,84] the key issue is to eliminate the traces of organic compounds that could have a detrimental effect on process baths and on

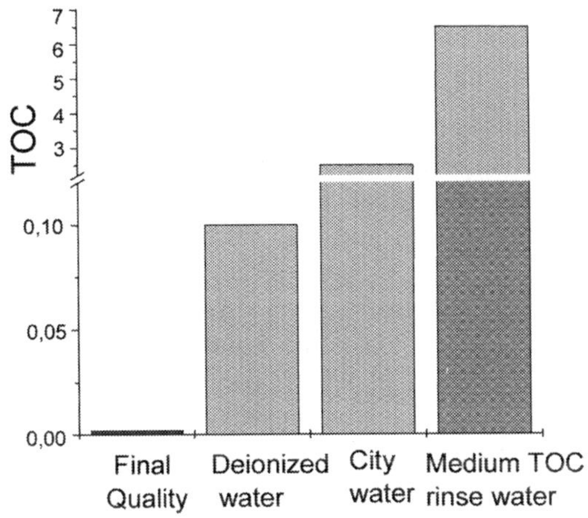

Figure 5 *TOC (Total Organic Carbon) change from city or rinse water to final quality required in electronic industry uses*[66]

the final products. Figure 5 shows this concept showing the level of TOC removal which should be reached with respect to some common situation of water quality.[82]

The suggested new water recycling process[84] is based on an advanced oxidation process (catalytic ozonization) followed by biological degradation and a final stage of further purification by a membrane. Ozonization in presence of a transition metal-supported catalyst allowed a 25% TOC removal with acceptable ozone doses (3-4 g O_3/g TOC). Filtration on a biological granular activated carbon (BGAC) then removed an additional 60% of the remaining TOC. With this combination, 70% of the initial TOC was removed, with extensive destruction of target compounds. To reach the ultimate target of 95% TOC removal, an additional treatment step by reverse osmosis (RO) thin-film composite (TFC) membranes is required. The TOC values in the RO permeate were typically lower than 0.20 mg/L, while the initial TOC values were around 7.5 mg/L. Such low TOC values could not be achieved when oxidation, BGAC filtration, or both were not used prior to the reverse osmosis step. Data were also validated in a pilot-scale prototype[84].

Although other types of production processes may have less demanding requirements, the development of water treatment technologies for its recycle requires generally large and costly research activity. The economic, management and environmental advantages in recycling the water can be incentives for companies having a more long term and strategic perspective of the environmental problem and water sustainability. However, the extensive recycle of water in industrial production requires the adoption of both incentives for research and legislative measures.

3 Conclusions

The increasing awareness on the issue of water sustainability has stimulated new interest in the development of advanced technologies for water treatment and in the use of solid catalysts to increase their performance, manageability, eco-sustainability and/or reduce their cost. Due to the large variety of wastewater situations in terms of composition and flow rate, a range of different and/or combination of different technologies, is required.

The use of solid catalysts is beginning to be investigated in relation to water and wastewater treatment. There are some areas such as photocatalysis, wet oxidation and selective reduction in which the use of solid catalysts is well established, but still significant new developments are expected from an improvement in both catalyst and reactor design. There are also other interesting areas in which investigation on the use of solid catalysts is at an earlier stage, such as in the catalytic ozonation and wet oxidation with hydrogen peroxide. Finally, there are new areas such as those where the use of unconventional sources of energy (waves, electronics) are used to generate radical or charged species at room temperature. The use of solid catalysts in conjunction with these unconventional sources of energy is at a very exploratory level, but results are interesting. A better understanding of the interaction between solid catalysts and radical/charged species in solution is required to allow really to make a step forward in the use of these novel catalytic technologies for water treatment.

In conclusion, the use of solid catalysts offers a range of potential advantages in water treatment and remediation technologies, but it is necessary to further develop background knowledge to fully understand limitations and advantages. Present results are promising and already a few industrial examples of applications exist. The next step requires a more rational design and engineering of catalysts and reactors, but this will open also new directions of research for catalysis and its use in solving world and environment global issues[21].

References

1. J.-M. Herrmann, *Catal. Today*, 1999, **53**, 115.
2. S. Malato, J. Blanco, A. Vidal, C. Richter, *Appl. Catal. B: Env.*, 2002, **37**, 1.
3. M. Kositzi, I. Poulios, S. Malato, J. Caceres, A. Campos, *Water Research*, 2004, **38**, 1147.
4. O. Alfano, D. Bahnemann, A. Cassano, R. Dillert, R. Goslich, *Catal Today*, 2000, **58**, 199.
5. A. Fujishima, T. Rao, D. Tryk, *J. Photochem Photobiol C: Photochem Rev.*, 2000, **1**, 1.
6. S. Chiron, A. Fernandez-Alba, A. Rodriguez, E. Garcia-Calvo, *Water Research*, 2000, **34**, 366.
7. M. Pera-Titus, V. Garcìa-Molina, M.A. Ba(241)os, J. Giméneza, S. Esplugas, *Appl. Catal. B: Env.*, 2004, **47**, 219.
8. R. Andreozzi, V. Caprio, A. Insola, R. Marotta, *Catal. Today*, 1999, **53**, 51.
9. (a) M. Anpo, *Stud. Surf. Sci. Catal.*, 2000, **130**, 157. (b) M. Anpo, M. Takeuchi, *J.*

Catal., 2003, **216**, 505. (c) M. Anpo, M. Takeuchi, K. Ikeue, S. Dohshi, *Current Opinion in Solid State & Materials Science*, 2002, **6**, 381.
10. R. Asahi, T. Morikawa, T. Ohwaki, K. Aoki, Y. Taga, *Science*, 2001, **293**, 269.
11. H. Kisch, W. Macyk, *ChemPhysChem*, 2002, 3, 399.
12. M. Stylidi, D.I. Kondarides, X.E. Verykios, *Appl. Catal. B: Env.*, 2003, **40**, 271.
13. H. Lachheb, E. Puzenat, A. Houas, M. Ksibi, E. Elaloui, C. Guillard, J.-M. Herrmann, *Appl. Catal. B: Env.*, 2002, **39**, 75.
14. F. Gracia, J.P. Holgado, A.R. Gonzalez-Elipe, *Langmuir*, 2004, **20**, 1688.
15. M. Noorjahan, V. Durga Kumari, M. Subrahmanyama, P. Boule, *Appl. Catal. B: Env.*, 2004, **47**, 209.
16. V. Keller, F. Garin, *Catalysis Communications*, 2003, **4**, 377.
17. N. Keller, V. Keller, F. Garin, M.J. Ledoux, *Materials Letters* , 2004, **58**, 970.
18. K.-H. Jung, S.R. Jang, R. Vittal, D. Kim, K.-J. Kim, *Bull. of the Korean Chemical Soc.*, 2003, **24**, 1501.
19. S. Sakthivel, H. Kisch, *Angewandte Chemie, Int. Ed.*, 2003, **42**, 4908.
20. B. Tryba, A.W. Morawski, M. Inagaki, *Appl. Catal., B: Env.*, 2003, **46**, 203.
21. G. Centi, P. Ciambelli, V. Palma, S. Perathoner, *Environmental Catalysis*, 2004, in press. Publisher: Kluwer Academic/Plenum Publishers, New York, N. Y
22. J. Levec, A. Pintar, *Catal. Today*, 1995, **24**, 51.
23. L. Oliviero, J. Barbier, Jr., D. Duprez, *Appl. Catal. B: Env.*, 2003, **40**, 163.
24. F. Luck, *Catal. Today*, 1999, 53, 81; Ibidem, 1996, **27**, 195.
25. Y. I. Matatov-Meytal, M. Sheintuch, *Ind. Eng. Chem. Res.*, 1998, **37**, 309.
26. D. Mantzavinos, M. Sahibzada, A.G. Livingston, I.S. Metcalfe, K. Hellgardt, *Catal. Today*, 1999, **53**, 93.
27. F.E. Hancock, *Catal. Today*, 1999, **53**, 3.
28. M. Campanati, G. Fornasari, A. Vaccari (Guest Ed.s), Natural Waters and Water Technology: Catalytic Science and Technology for Water, *Catal. Today*, 1999, **53(1)**, 158 pp.
29. J.B. Beziat, M. Besson, P. Gallezot, S. Durecu, *J. Catal.*, 1999, **182**, 129.
30. L. Oliviero, J. Barbier, D. Duprez, *Appl. Catal., B: Env.*, 2003, **40**, 163.
31. H.S. Joglekar, S.D. Samant, J.B. Joshi, *Water Research*, 1991, **25**, 135.
32. B. Kasprzyk-Hordern, M. Zi(243)Å*ek, J. Nawrocki, Appl. Catal. B: Env.*, 2003, **46**, 639.
33. B. Legube, N. Karpel Vel Leitner, *Catal. Today*, 1999, **53**, 61.
34. M.A. Tarr, *Environmental Science and Pollution Control Series*, 2003, **26**(Chemical Degradation Methods for Wastes and Pollutants), 165.
35. E. Neyens, J. Baeyens, *J. of Hazardous Materials*, 2003, **98**, 33.
36. C. Merli, E. Petrucci, A. Da Pozzo, M. Pernetti, *Annali di Chimica* (Rome, Italy), 2003, **93**, 761.
37. A. Safarzadeh-Amiri, J.R. Bolton, S.R. Cater, *J. of Adv. Oxidation Techn.*, 1996, **1**, 18.
38. R.J. Bigda, *Chem. Eng. Progress*, 1995, **91**, 62.
39. S.J. Masten, S.H.R. Davies, *Adv. in Env. Science and Techn.*, 1994, **28**(Environmental Oxidants), 517.
40. H. Fenton, *J. Chem. Soc.*, 1894, **65**, 899.
41. (a) G. Classics, S. Visentin, C. Cavagnis, *Riv. It. Sost. Gr.*, 1994, **71**, 561. (b) (b) C. Collivignarelli, V. Riganti, S. Teruggi, F. Montemagno, *Ingegneria Ambientale*,1997, **26**, 332. (c) C. Collivignarelli, M. Baldi, G. Bertanza, G. Bissolotti, R. Calvi, M. Del Rio, M. Pergetti, *Ingegneria Ambientale*, 1997, **26**, 109.
42. A.I. Balcioglu, A.I. Alaton, M. Oetker, R. Bahar, N. Bakar, M. Ikiz, *J. Env. Science*

and Health, Part A: Toxic/Hazardous Substances & Env. Eng., 2003, **A38**, 1587.
43. I. Sekoulov, Application of hydrogen peroxide (H_2O_2) in waste water technology., DECHEMA Monographien, 1977, **81**, 243.
44. G. Centi, S. Perathoner, T. Torre, M.G. Verduna, *Catal. Today*, 2000, **55**, 61.
45. G. Centi, S. Perathoner, *Topics in Catalysis*, 2004, submitted.
46. G. Centi, S. Perathoner, *Appl. Catal. B: Env.*, 2003, **41**, 15.
47. A.J. Lecloux, *Catal. Today*, 1999, **53**, 23.
48. K.D. Vorlop, T. Tacke, *Chemie Ingenieur Technik*, 1989, **61**, 836.
49. (a) U. Prusse, M. Hahnlein, J. Daum, K.D. Vorlop, *Catal. Today*, 2000, **55**, 79. (b) K.D. Vorlop, U. Prusse, in *Catalytic Science Series*, 1999, **1**(Environmental Catalysis), 195.
50. R. Dittmeyer, V. Hollein, K. Daub, *J. Mol. Catal. A: Chem.*, 2001, **173**, 135.
51. N. Munakata, M. Reinhard, in *Physicochemical Groundwater Remediation*, A.J. Smith, S.E. Burns, Editor(s): 2001, p. 45. Publisher: Kluwer Academic/Plenum Publishers, New York, N. Y
52. C. Schüth, S. Disser, F. Schüth, M. Reinhard, *Appl. Catal. B: Env.*, 2000, **28**, 147.
53. G. Centi, *J. Molec. Catal. A: Chem.*, 2001, **173**, 287.
54. M. Marella, M. Tomaselli, F. Gerolin, R. Dittmeyer, K. Daub, G. Strukul, G. Centi, P. Ruiz Barrientos, F. Luck, M. Van Donk, Maarten, Catalytic contactor and process to remove nitrate, nitrite, perchlorate and other harmful compounds from contaminated water with hydrogen, *Ger. Offen.* (*Patent*), 2002, DE 10064622 (assigned to Venezia Tecnologie S.p.A., Italy; G.I.E. Anjou-Recherche; CH2MHill, San Fernando de Hanares).
55. (a) J.L. Vogan, R.M. Focht, D.K. Clark, S.L. Graham, *J. Hazardous Materials*, 1999, **68**, 97. (b) A.R. Gavaskar, *J.Hazardous Materials*, 1999, **68**, 41.
56. H.C. Fry, K.D. Karlin, P.G. Hoertz, D. Wong-Verelle, G.J.J. Meyer, *Preprints of Extended Abstracts presented at the ACS National Meeting*, American Chemical Society, Division of Environmental Chemistry, 2002, **42**, 307.
57. G. Centi, N. Grande, S. Perathoner, *Catal. Today*, 2002, **75**, 69.
58. P.E. Savage, *Chem. Rev.*, 1999, **99**, 603.
59. S.N.V.K. Aki, M.A. Abraham, *Environ. Prog.*, 1998, **17**, 246.
60. (a) P.E. Savage, S. Gopalan, T.I. Mizan, C.J. Martino, E.E. Brock, AIChE J., 1995, 41, 1723. (b) J. Yu, P.E. Savage, *Appl. Catal. B: Env.*, 2001, **31**, 123.
61. U. Armbruster, A. Martin, A. Krepel, *Appl. Catal. B: Env.*, 2001, **31**, 263.
62. (a) T.M. Olson, P.F. Barbier, *Water Research*, 1994, **28**, 1383. (b) I. Hua, M.R. Hoffmann, *Environ. Sci. Technol.*, 1997, **31**, 2237. (c) Y. Ku, K. Chen, K. Lee, *Water Research*, 1997, **31**, 929. (d) K. Vinodgopal, J. Peller, O. Makogon, P. Kamat, *Water Research*, 1998, **32**, 3646.
63. C. Petrier, M. Lamy, A. Francony, A. Benahcene, B. David, *J. Phys. Chem.*, 1994, **98**, 10514.
64. J. Ge, J. Qu, *Appl. Catal. B: Env.*, 2004, **47**, 133.
65. N.H. Ince, G. Tezcanli, R.K. Belen, PI.G. Apikyan, *Appl. Catal. B: Env.*, 2001, **29**, 167.
66. E.P. Reddy, L. Davydov, P.G. Smirniotis, *Appl. Catal. B: Env.*, 2003, **42**, 1.
67. L. Davydov, E.P. Reddy, P. France, P.G. Smirniotis, *Appl. Catal. B: Env.*, 2001, **32**, 1.
68. K.P. Francke, H. Miessner, R. Rudolph, R., *Catal. Today*, 2000, **59**, 411.
69. M.A. Malik, S.A. Malik, *Chinese J. Chem. Eng.*, 1999, **7**, 351.
70. Z. Chen, V.K. Mathur, *Ind. & Eng. Chem. Research*, 2002, **41**, 2082.
71. A. Wekhof, W. Gudernatsch, *Chemie-Technik* (*Heidelberg*), 2000, **29**, 26.

72. (a) M.A. Malik, A. Ghaffar, S.A. Malik, *Plasma Sources Science & Techn.*, 2001, **10**, 82.(b) M.A. Malik, *Plasma Sources Science & Techn.*, 2003, **12**, S26.
73. Y. Shimizu, Y. Nakamoto, T. Hyodo, M. Egashira, *Electrochemistry* (Tokyo, Japan), 2004, **72**, 92.
74. (a) A.K. Pikaev, *High Energy Chem.*, 1994, **28**, 1. (b) P. Gehringer, H. Eschweiler, *Ozone Science and Eng.*, 1999, **21**, 523. (c) J.R. Bolton, J.E. Valladares, J.P. Zanin, W.J. Cooper, M.G. Nickelsen, D.C. Kajdi, T.D. Waite, C.N. Kurucz, 1998, *J. Advan. Oxid. Technol.* **3**, 174.
75. G. Centi, S. Perathoner, *Catal. Today*, 2003, **77**, 287.
76. G. Centi, M. Gotti, S. Perathoner, F. Pinna, *Catal. Today*, 2000, **55**, 51.
77. M. Sheintuch, Y.I. Matatov-Meytal, *Catal. Today*, 1999, **53**, 73.
78. G. Centi, S. Perathoner, *Catal. Today*, 2003, **79-80**, 3.
79. (a) Sustainable use of Europe's water?, S.C. Nixon, T.J. Lack, D.T.E. Hunt, C. Lallana, A.F. Boschet, Environmental assessment series No 7, European Environment Agency, Copenhagen, 1999. (B) Water resources problems in Southern Europe – An overview report, EEA: Topic report 15/1997, Inland Waters, European Environment Agency, Copenhagen, 1997.
80. G. Centi, S. Perathoner, Sustainable water use and technologies, *La Chimica e L'Industria* (*Milan*), 2001, **83**, 43.
81. Compendium of waste water treatment and water purification technologies, C. Buitrago, G. Centi, A. Lodolo, S. Miertus, ICS Pub.: Trieste 2002.
82. G. Centi, S. Perathoner, *Catal. Today*, 1999, **53**, 11.
83. (a) M. Djafer, F. Luck, J.P. Rose, D. Cretenot, *Water Science and Technology*, 2000, **41**(8, Sludge Management for the 21st Century), 77. (b) J.P. Rose, C. Bonnin, L. Patria, D. Cretenot, F. Luck, FIn Proceedings WEFTEC 2000, Annual Conference & Exposition on Water Quality and Wastewater Treatment, 73rd, Anaheim, CA, United States, Oct. 14-18, 2000, 219.
84. (a) M. Djafer, F. Luck, R. Toupin, *Industrial Wastewater*, 1999, 7, 25. (b) M. Djafer, F. Luck, R. Toupin, in Proceedings of the WEF/Purdue Industrial Wastes Conference, Indianapolis, IN, United States, June 27-30, 1999, 840. (c) F. Luck, M. Djafer, N. Karpel, V. Leitner, B. Gombert, B. Legube, *Water Science and Technology*, 1997, **35**(4, Oxidation Technologies for Water and Wastewater Treatment), 287.

3

Novel Supported Metal Oxide Adsorbents and Catalysts Prepared from Polynuclear Metal Complexes

BY MARK G. WHITE

1 Introduction

The literature was reviewed to describe the newest efforts to synthesize and characterize supported polynuclear metal complexes as adsorbents and catalysts. This review includes our attempts to model the equilibrium structures and properties of the metal complexes, using simple quantum mechanics, as a means to understand better the interactions between the surface and the metal complexes. Special attention is directed towards the characterization of the supported metal complexes before and after ligand removal. We compare these modeling results with observations in the literature so as to understand better the fundamental processes that govern the interactions between the metal complexes and the surfaces. With this enhanced understanding of these governing factors, it should be easier to prepare oxide solids decorated with metal complexes having the desired physico-chemical properties.

2 Background

Chemists have prepared metal complexes containing metal atoms/ions as a means to understand better the structure, chemical bonding, and properties of metals and metal ions. One of the first efforts to affix these metal complexes to a surface as a means to create a supported catalyst was reported by Ballard[1] followed by reports collected by Yermakov[2], *et al.* and Basset[3], *et al.* We distinguish here between metal complexes that contain zero-valent metals and those that show metal cations and we limit this review to complexes containing metal ions as others have published extensive reviews of zero-valent, metal clusters and their chemistry.[4,5,6] In our previous three reviews on the chemistry of supported, polynuclear metal complexes, we described efforts to synthesize and characterize oxide-supported, metal complexes as adsorbents, catalysts and precursors to supported metal oxides.[7,8,9] In one application of this technology, efforts were

Catalysis, Volume 18

directed towards decorating high-surface area oxide surfaces with a monolayer of metal complexes that were shown to be active for adsorption and/or reaction. For another application of this technology, where the metal complexes were not active owing to the complete coordination of the metal ions, the solids were heated to decompose the metal complexes so as to make highly dispersed adsorbents and catalysts that resist sintering at high temperature in oxidizing and reducing conditions. This latter application of the technology may find general use in preparing supported, mixed-metal oxides as adsorbents and catalysts. In theory, this technique can be applied to a wide variety of oxide supports that have the proper surface chemistry described below.

3 The Chemistry of Oxide Surfaces

Before we describe the chemistry of interactions between the support and the polynuclear metal complexes, it is appropriate to describe some features common to many metal oxide supports. Some metal oxides that are used as supports show hydroxyl groups and bridging oxygen atoms at the oxide surface (Figure 1).

These hydroxyl groups may be isolated (**I**) or bridging between two metal cations (**II**). Species (**II**) is often observed in stable metal complexes where the oxygen atoms experience both covalent and coordination bonding to the metal cations. Often, two hydroxyl groups may appear on the same metal cation and may experience hydrogen bonding (**III**). The surface may also show metal oxanes (**IV**), while other oxides show cation and anion defect sites. While some metal oxides may have only a subset of these surface features, others, such as alumina, may show a rich diversity of all of these features.[10] For some metal oxide supports, such as silica, the population of these surface features depends upon the temperature and water partial pressure of the gas in contact with the solid. It is well known that the population of silanols and siloxanes are governed by the equilibrium shown below:

$$2[\text{-SiOH}] + \Delta \Leftrightarrow \text{-Si-O-Si-} + H_2O\ (g)$$

A priori knowledge of this equilibrium may play a role in the design of the materials synthesis when it is known that either the surface silanols or the surface siloxanes are the reactive species to anchor the metal complexes to the surface.

The surface chemistry of these features must also be acknowledged as a factor that influences the interaction between the support and the polynuclear metal

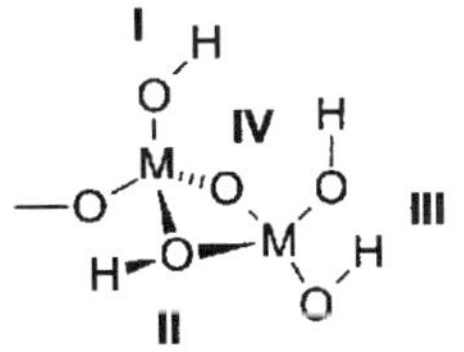

Figure 1 *Schematic diagram of the surface of a metal oxide*

complexes. For example, the OH groups may be acidic (labile proton), basic (labile OH^-), and neutral. Indeed, it is the knowledge of the chemistry of the OH group that allows one to design a metal complex that can be anchored to the surface. Consider as an example, the interaction between a simple metal complex, copper(II) acetylacetonate, which can react with either an acidic OH group or a basic OH group.[11]

Acidic surface proton:

$$Cu(acac)_2 + H^{+\ 1-}O\text{-}M\text{-} \rightarrow H(acac) + (acac)\text{-}Cu\text{-}O\text{-}M\text{-}$$

Basic surface OH^-:

$$2Cu(acac)_2 + HO^{1-\ +}M\text{-} + 2ROH \rightarrow (acac)\text{-}Cu\text{-}(\mu OR)_2\text{-}Cu(acac) + 2H(acac) + HO^{1-\ +}M\text{-}$$

When acidic, surface OH groups are present, the copper acetylacetonate reacts with the proton to form the stable pentanedione molecule and to form a surface-grafted Cu(acac) through the metal oxide. This acid-catalyzed reaction will occur in the appropriate solvent, such as acetonitrile. Basic surface OH groups are a catalyst in the presence of an alcohol solvent to form the μ-methoxy, dinuclear copper acetylacetonate complex and two molecules of pentanedione.[11] This dinuclear copper complex is readily recognized by its deep purple color as opposed to the vivid blue color of the mononuclear copper complex and this dinuclear metal complex can be formed readily on the surface of MgO when methanol is the solvent[11]. This simple illustration is but one example of a general acid/base reaction that is reported in a larger body of findings.[7-9,12] If acidic and basic OH groups are present on the same surface, then it becomes apparent that grafted mononuclear and dinuclear copper complexes may be the result providing that a simple alcohol, such as methanol, is the solvent. One may preclude the formation of the Cu dimer complex by using another solvent, such as acetonitrile, which is non-coordinating to the copper complexes. This system is illustrative of the care that must be exercised when designing a synthesis to prepare a 'homogeneous' surface having the desired site chemistry.

The choice of support not only includes a specification for the desired physical properties such as surface area, pore volume, *etc*, otherwise known as surface texture, but this choice also must include a consideration of the surface features likely to be present under the conditions used for the synthesis of the supported metal complex and their intrinsic chemistry. These surface features determine in part the strategy that one uses in the design and the synthesis of the polynuclear metal complex. This strategy may also include the restructuring of the surface so as to inculcate into the surface the desired chemistry. This restructuring has been practiced by those who have functionalized the surface of a high surface area support to prepare a partitioning column for chromatography. Under the moniker of self-assembled surfaces, the scientists have employed simple reactions to change the chemistry of metal surfaces such as gold or silver into a surface having organic molecules terminating with amine, mercaptan, or carboxylic acid groups to act as sensors in plasmon resonance spectrometers.[13] More recently, researchers are restructuring the surface of silica with the goal to make 'single-site' polymerization catalysts[14]. Thus, it is left to the imagination of the catalytic scientist to 'engineer' the desired surface by exploiting these simple ideas of chemistry.

4 The Chemistry of Decorating Oxide Surfaces with Metal Complexes – Liquid Phase Technique

Over the past twenty years we have developed a design heuristic for the preparation of supported, mixed metal oxides. This approach recognizes the problems encountered when using water as a solvent for the salts of transition metal ions. Many of these metal ions show a facile equilibrium between their oxidation states that show different structures which form at different values of the solution pH. As an illustration of this problem consider the task of synthesizing a mixed metal oxide of Cu/Cr. We abandoned our efforts to prepare a supported Cu/Cr catalyst from the basic, ammonium salts having an unique structure when we showed that solids could be precipitated from solution as single crystals having different, cubane-type structures and having different Cu/Cr ratios. These solids could be developed from aqueous solutions by changing only slightly, the pH of the solution[15]. This frustrating attempt led us to abandon water as a solvent and to use aprotic, non-coordinating solvents that do not compete with cationic metal complexes for ion exchange with surface protons and will not coordinate to metal complexes. One such solvent is acetonitrile and it is especially good as a solvent for perchlorate salts of metal complexes. While one finds alcohols[16] and arenes[17] frequently used as solvents for metal complexes, we believe that acetonitrile can be used with excellent results with cationic metal complexes when the goal is to prepare a surface decorated with replicas of a metal complex.

We have been successful in decorating surfaces with neutral, metal complexes, $(M^{m+})_n(L^{n-})_m$ having anionic ligands, L^{n-} that can be removed at low temperature by the reaction with surface protons to produce neutral species, H_nL^{n-}, by the following reaction:

$$(M^{m+})_n(L^{n-})_m + (H^+)(^-O\text{-}Su) \rightarrow (M^{m+})_n(L^{n-})_{(m/n-1)}OSu + H_nL^{n-}$$

This reaction is a generalized form of the reaction we illustrated earlier between copper acetylacetonate and the surface protons of silica. Often only a subset of the ligands reacts with surface protons to form coordination bonds with surface oxides (SuO): $(M^{m+})_n(L^{n-})_{(m-1)}OSu$. The remaining ligands can be removed by careful heating in either oxidizing or neutral atmospheres. Examples of ligands, L^{n-} that we have used in synthesizing polynuclear metal complexes that are suitable for decorating the surfaces of oxides include acetylacetonate, N-N-dialkylalkoxyamine, ethylenediamine, and others. The acetylacetonate anion is particularly attractive owing to the fact that it combines with most of the metals in the Periodic Table. Such synthetic flexibility makes it possible to prepare a large number of supported metal oxide catalysts using the metal acetylacetonate as the precursor.

Alternatively, one can prepare supported metal oxides by beginning with cationic metal complexes $[M_1M_2L]^{m+}A^{n-}$ for which the anion, A^{n-}, combines with the surface protons as the cationic metal complex forms an ionic bond with surface anion sites: $^{-}$OSu. By this approach the ligands may remain attached to cation complex. Examples of the ligands that we have used include N-N-dialkylalkoxyamine, and ethylenediamine. We showed how supported, mixed, metal oxide could be prepared by starting with a polynuclear metal complex that was a perchlorate salt. The same metal complex could be synthesized as the PF_6^- salt to preclude the 'explosive' nature of perchlorate salts. One polynuclear metal complex, $\{[O\text{-}CH_2CH_2\text{-}NR_2\text{-}Cu\text{-}\mu OH\text{-}]_6\text{-}M\}^{3+}(ClO_4)_3$ where M = Al, Cr, or Fe, was supported on silica as a monolayer.[18] This metal complex showed both Lewis and Bronsted acid sites, the strength of which could be systematically changed by changing the metal ion, M.[19] When the ligands, $O\text{-}CH_2CH_2\text{-}NR_2$, were removed, the remaining metal oxide was not acidic and the metals sintered easily. These systems are examples where the ligands have basic nitrogen's and/or oxygen's to facilitate activation of surface OH into surface oxides + protons. When the metal complex show ligands that which are not sufficiently basic, such as ethylenediamine, the activation of surface OH into protons can be facilitated with the use of a co-solvent which is a non-coordinating base, such as trialkyl amine. This amine forms protonated trialkylamine as intermediate to facilitate activation of the surface proton which then reacts with the anionic ligand of the metal complex to facilitate its removal as a neutral molecule.

In these applications, metal complexes were designed to have the desired constitution and site geometry and these were attached to a surface. While this approach is appealing, the cost of preparing large quantities of these species may be very expensive and the corresponding improvement in yields may not warrant the increased cost. Therefore, one may find it necessary to approach the problem of preparing highly selective and active catalysts in a different way. Very recently, Coulier, *et al.* showed some interesting ways to modify the chemistry of transition metal ions supported on alumina by selective chelation with agents such as ethylenediamine (en), ethylenediaminetetracetic acid (edta), *etc.*[20] They showed that the sulfidation of Co is retarded while that of Mo is facilitated when the surface of the model Co/Mo/alumina catalysts was treated with (edta). This result was explained easily in terms of the low stability of Mo-edta for ligand exchange; whereas, the Co-edta is very stable and does not readily engage in ligand exchange for the sulfide ligand. As a result, much of the surface Mo ions were preferentially sulfided. These modified model catalysts were shown to have the highest activity for a hydrodesulfurization (HDS) probe reaction: thiophene HDS. A similar result was reported for the modification and subsequent characterization of a silica-supported Ni/Mo catalyst modified with (edta).[21] Thus, the selective, post-synthesis modification of surface metal cations by certain stable organic molecules, may be a general way to produce highly selective, supported metal oxide catalysts without the very high cost associated with making the metal complex from the very beginning provided that the surface-bound organic molecules have a reasonable 'life' on the catalyst.

5 The Chemistry of Decorating Oxide Surfaces with Metal Complexes – Gas Phase Technique

Metal oxide surfaces can be decorated with certain metal complexes that can be vaporized at conditions below those required for the decomposition of the complex. A group in Finland has published extensively on this technique which they call Atomic Layer Epitaxy.[22,23] One group of metal complexes, the metal acetylacetonates, are often used as the sources of the volatile metal compounds as they can be easily volatilized into a flowing inert gas stream at modest temperatures. The fluoro-form of the metal acetylacetonates, *e.g.*, hexafluoroacetylacetonate, are often used owing to their high volatility under partial vacuum conditions. The chemistry of the reaction between the gaseous metal complexes and the surfaces has been reported by a number of groups.[24, 25]

As with the liquid phase reaction, one mechanism involves the reaction of the acac ligands with the acidic, surface protons by the following reaction:

$$M^{n+}(acac)_n + H^{+\,-}OSu \rightarrow Su\text{-}O\text{-}M^{n+}(acac)_{n-1} + H(acac)$$

This reaction may continue past the first ligand exchange reaction to completion. When the surface of metal oxide also contains coordinately unsaturated sites, such as on the surface of some aluminas, then another reaction is possible between the volatile metal acetylacetonate and the surface:

$$2(Al^{3+})_{surface} + M^{n+}(acac)_n \rightarrow [Al(acac)_x]^{(3-x)}{}_{surface} + Al(acac)_3 + [M^{n+}]_{surface}$$

and the pentanedione product of the reaction with surface protons may also react with surface alumina ions as shown below:

$$2(Al^{3+})_{surface} + (3+x)H(acac) \rightarrow [Al(acac)_x]^{(3-x)}{}_{surface} + Al(acac)_3 + (3+x)[H]_{surface}$$

This science may be applied to the technology of catalyst synthesis especially when the metal complexes are too stable to be activated at the mild reaction conditions imposed by the liquid phase procedure. For example, we[11, 26] reported that $Cr(acac)_3$ would not react with the surface of silica when the system was refluxed in acetonitrile at 1 atm owing to the high, intrinsic stable of the *tris*-$Cr(III)(acac)_3$. However, others[27,28,29] showed that high loadings of Cr/support could be obtained using the gas phase technique when starting with $Cr(acac)_3$. The energy required to active the Cr-(acac) bond was provided by the thermal energy of the unsupported $Cr(acac)_3$ molecules. Thus, for certain applications, the gas-phase technique can provide a synthetic pathway to the synthesis of supported metal oxide catalysts that cannot be accessed through the lower-temperature, liquid phase technology.

One may apply the gas-phase technique to make large amounts of supported metal oxides by the following procedure. The powder of a metal oxide support may be confined as a fixed bed with the gaseous metal acetylacetonate (sublimed from a source at 180-200°C) transported through the bed by an inert carrier gas at a temperature sufficiently high to cause the reaction (>250°C). The extent of reaction between the metal source and the metal oxide support can be determined by monitoring the composition of pentanedione in the gas phase. The concentration of pentanedione falls to zero in the effluent gas stream when the metal oxide bed is 'saturated' with the source metal. The conditions can be optimized for producing the desired weight loading using the usual reactor control parameters of space-time, concentration of metal complexes, and temperature. Alternatively, a more uniformly decorated catalyst may be produced using a fluidized bed of the support oxide provided that the particle sizes are sufficiently small to allow fluidization.

6 Quantum Mechanical Modeling – Equilibrium Structures of Isolated Metal Complexes

Some success has been reported on modeling the properties of metal complexes using quantum mechanical tools. The level of modeling ranges from semi-empirical techniques such as PM3 to *ab initio* methods.

6.1 VO(acac)₂. – An *ab initio* calculation was completed to determine the equilibrium geometry of this metal complex.[30] For this calculation and our subsequent calculations, the acetylacetonate ligand was approximated by the (acac') ligand which shows the methyl groups replaced by hydrogens so as to decrease the time required to complete the calculations. This simplification has been found satisfactory in previous studies[31].The predicted equilibrium structure shows the (acac') ligands were bent away from a plane that is perpendicular to the V=O bond (Fig. 2).

The bent (acac) ligands were attributed to the influence of the axial oxygen atom (O5). The observed angle between the two ligand planes is 163° by XRD[32] and it is 145° by electron diffraction[33]. The calculated geometries show this angle to be 159-166° depending upon the model using in the calculations. Thus, the predicted equilibrium structure appears to be consistent with the observations.

6.2 Cu(acac)₂. – Semi-empirical (PM3) and *ab initio* calculations (DFT) were completed for the 'isolated' Cu(acac')₂. These calculations produced very similar equilibrium structures where Cu and the two (acac') ligands are in one plane (Fig. 3).

The structure determined by XRD of crystalline Cu(acac)₂ shows a slightly bent structure (Figure 4) where the ring carbon atoms are at an angle of 13°

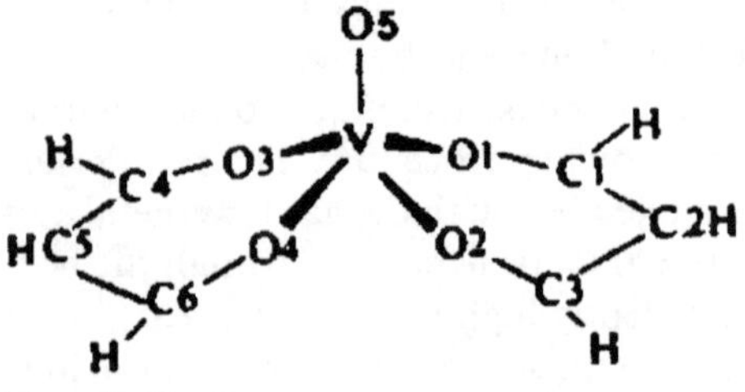

Figure 2 *Diagram of (acac')₂V=O*

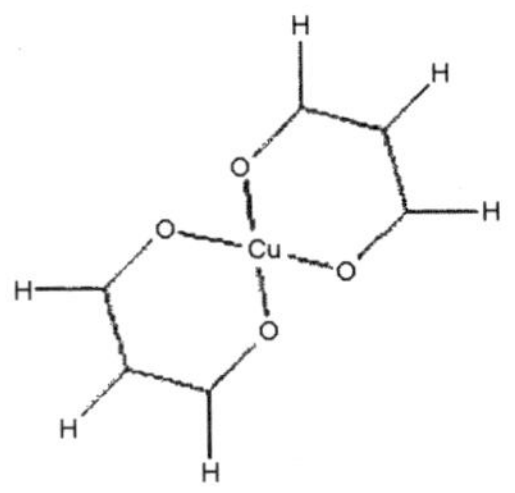

Figure 3 *Equilibrium Geometry Predicted by DFT and PM3*

Figure 4 *ORTEP of Cu(acac)$_2$*

relative to the plane described by the copper atom and the four equatorial oxygen atoms.[34] Thus, we concludethat the higher level calculations cannot predict the bent structure observed in the crystalline Cu(acac)$_2$ and that the small structural subtleties of a metal complex in a crystal are perhaps the effects of atoms in neighboring metal complexes and these interactions with the neighbor appear to be a factor in determining the geometry of the complex.

Powder XRD data for polycrystalline Cu(acac)$_2$ show that the metal complexes arrange in a 'herringbone' fashion (Figure 5). With this arrangement, the (acac) ligand of one metal complex is brought into the coordination sphere of a copper from a neighboring metal complex to become an axial ligand. This type of interaction could be responsible for the slightly curved structure of the (acac) ligands. Data from electron paramagnetic studies of Cu(benzac)$_2$ diluted in Pd(benzac)$_2$ suggest strongly that the ligands interact with the metal centers via the axial field, and when the concentration of Cu^{2+} ions is sufficiently high, then Cu-Cu coupling can occur to influence the EPR spectrum.[35] Unfortunately, much larger ensembles of atoms cannot be modeled by the higher-level, *ab initio* programs.

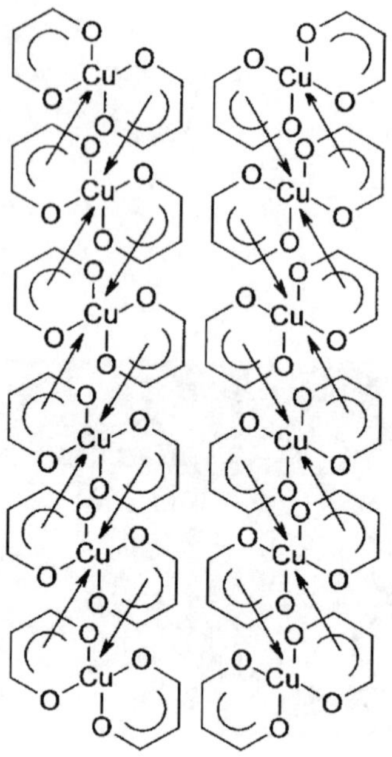

Figure 5 *'Herringbone' arrangement of $Cu(acac)_2$*

7 Quantum Mechanical Modeling– Properties of Isolated Metal Complexes

7.1 $VO(acac)_2$. – The predicted vibrational spectrum of this molecule is shown in Table 1 along with the observed spectrum[31].

The predicted vibrational frequencies of the $VO(acac)_2$ metal complex are higher than the observed frequencies in this region of the IR spectrum. However, notice that the overtone vibration of the γ(C–H) is not present in either the observed or the predicted spectrum for this molecule. The predicted and observed equilibrium structures show bent (acac) ligands, and we conjecture that the absence of their overtone in the IR spectra for this molecules is a result of the bending of the (acac) ligands away from a plane perpendicular to the V=O bond that includes the V atom. Therefore, we will attempt to apply the results observed for this molecule regarding the shape of the ligands and the IR spectrum to understanding better the IR spectrum of supported copper acetylacetonate.

The entries in Table 1 for $Cu(acac)_2$ have been assigned in the literature and we used these assignments to interpret the effect of the interaction with a hydrophilic surface such as silica.[27] The 2[γ-(C–H)] vibration is observed at 1552 cm^{-1} in the IR spectrum of polycrystalline $Cu(acac)_2$ and in the IR spectrum of a silica-supported metal complex that has been partially decomposed by heating so as to remove only one (acac) ligand so that the stoichiometry of the resulting supported metal complex was Cu(acac). Thermal decomposition of the supported metal complex in a TGA confirmed that the (acac) ligands were removed

Table 1 *IR Vibrations of Acetylacetonate Metal Complexes*

assignment	*(acac)$_2$V=O* *observed*	*predicted*	*(acac)$_2$Cu* *observed*	*(acac)$_2$Cu/silica* *observed*	*(acac)Cu/silica* *observed*
ν_s(C–O)	1600	1756	1576	1580	1590
ν_{as}(C–O)	1561	1737	1420	1441	1440
2[γ-(C–H)]			1552	–	1552
ν_{as}(C–C–C)	1523	1635	1532	1529	1520
ν_s(C–C–C)	1287	1574	1275	1281	–
δ_s(C–H)	1187	1240	1356	1370	1390

sequentially in two steps[11]. A lower temperature range of 200-290°C was required to remove the first ligand, and a higher temperature range (300-380°C) was required to remove the second ligand.[7,8,11] Thus, it was possible to determine the temperature required to remove only one (acac) ligand and so prepare a sample that still has only a single ligand. These findings were confirmed by others for the Cu^{2+} system[8] and for other systems.[12] On the other hand, the same sample before thermolysis did not show a vibration at this frequency (1552 cm^{-1}) and from the elemental analysis of this supported metal complex we were certain that the metal to ligand stoichiometry was consistent with its formula: Cu(acac)$_2$. Earlier, we hypothesized that the difference between the spectra of the supported and unsupported Cu(acac)$_2$ was due to hydrogen bonding between the support and the (acac) ligands of the complex.[11, 27] However, subsequently, we dismissed this hydrogen-bonding to ligand hypothesis and suggested that proximity of the metal complex to the surface was responsible for this effect without describing in detail how this effect was responsible.[36] We now use molecular modeling of the supported metal complex to show that steric hindrance of the ligand with surface siloxides may explain our observations.

8 Quantum Mechanical Modeling – Metal Complexes Decorating the Oxide Surface

8.1 Cu(acac)$_2$/silica. – We attempted to model copper acetylacetonate decorating the surface of silica using the semi-empirical program PM3 offered in the Spartan-02 package. The first step was to select a model for the silica surface. The challenge here is to choose a 'basis set' of silicon and oxygen atoms to build a structure that simulates the amorphous silica support developed from flame hydrolysis of silicon tetrachloride but also to choose a 'basis set' of atoms sufficiently small to permit modeling by the quantum mechanics program. For this purpose we arranged 10 silicon atoms (Figure 6) showing tetrahedral coordination to bridging oxides (initially having bond angles Si-O-Si = 180°). This model was constructed to show silicon atoms having silicon coordination numbers of 2 and 3 (Q^2 and Q^3, respectively) but no silicon atoms demonstrated coordination to 4 other silicon atoms (Q^4) through bridging oxygen atoms owing to the small size of the nanostructure (d ~ 1 nm). Copper acetylacetonate was

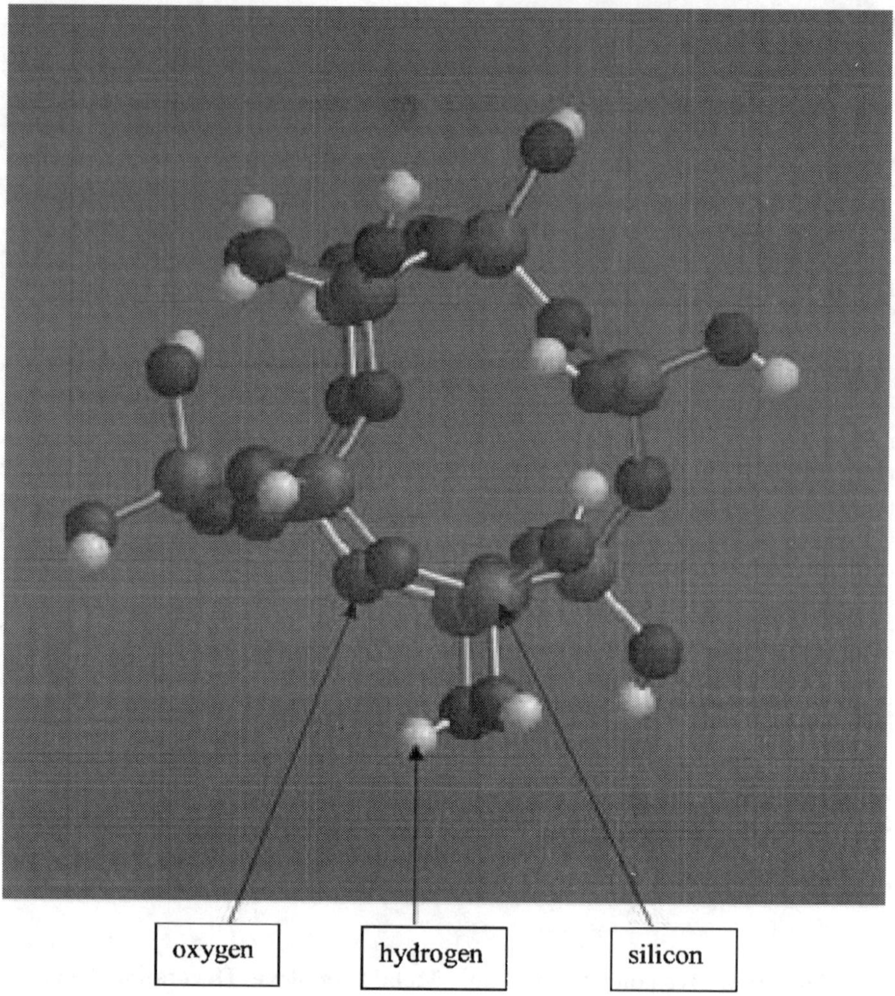

Figure 6 *Model of silica nanoparticle*

modeled (Figure 3) using the same software but the methyl groups were replaced with hydrogen's so as to reduce the number of atoms that must be included in the model of the supported metal complex.

These two models of Cu(acac')$_2$ and $(SiO_2)_{10}$ were subjected to an energy minimization to define the equilibrium structure of the combination beginning from different initial positions. The equilibrium geometry predicted by a PM3 energy minimization is shown in Figure 7.

Notice how the two ligands are bent away from the silica surface. Also, notice how the Cu cation is positioned over the silanol oxygen and how the proton of the silanol is positioned between the oxygen atoms of the two (acac') ligands. This bent structure for the Cu(acac')$_2$ decorating a silica surface is similar to the

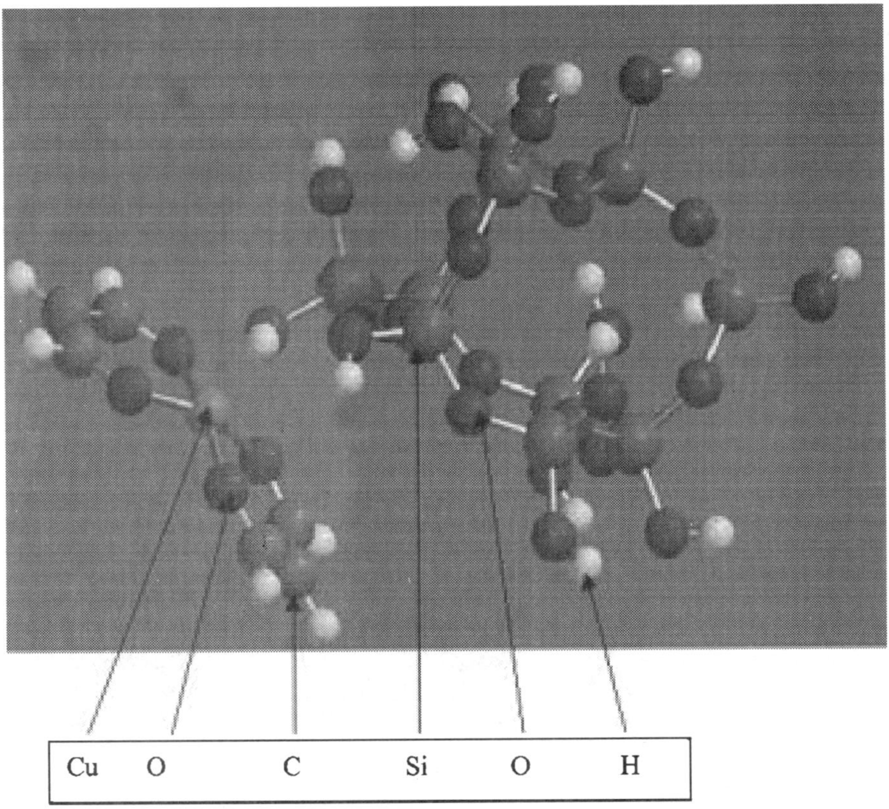

Figure 7 *Predicted Equilibrium Structure of Cu(acac)$_2$/silica*

bent structure we showed earlier for $(acac)_2V=O$. The simulation of the vanadyl acetylacetonate suggested that the electron density in the $V=O$ repelled the π-electron rich system of the (acac') ligands. We suggest that this same electron-electron repulsion could be operative to bend the (acac') ligands away from that part of the silica surface not decorated with protons while the hydrogen-bonding of the surface protons to the acetylacetonate oxygens is the driving force to attract the complex to the surface.

We simulated thermal decomposition of the supported Cu(acac)$_2$ by removing one of the ligands and the surface silanol proton, then forming a bond between the Cu cation and the surface siloxide. This structure was subjected to the PM3 energy minimization to produce the equilibrium geometry shown in Figure 8.

Notice how the remaining (acac') ligand tilts away from the surface and this ligand is 'flat' in that all atoms of the ligand define a plane. The experimental data of IR for the thermal decomposition of a supported Cu(acac)$_2$ complex shows the return of the 1552 cm^{-1} vibration which suggests that the remaining (acac) ligand is not in intimate contact with the silica surface.[11] If we also use the predictions to interpret further the IR data, we may suggest here that a flat (acac) ligand will

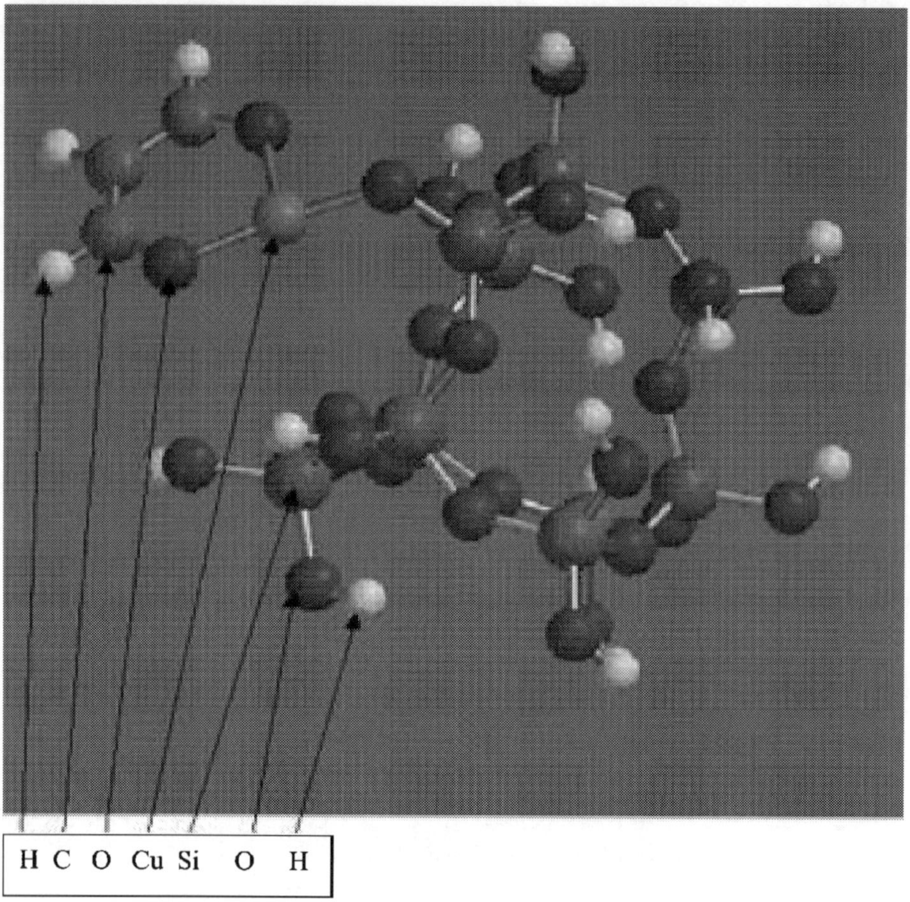

Figure 8 *Predicted Equilibrium Structure of Cu(acac) on silica after partial thermolysis.*

produce the 1552 cm^{-1} vibration. This simple model appears to explain well the observed IR data for $Cu(acac)_2$ placed on silica as a monolayer and the data for the thermolysis of $Cu(acac)_2 \rightarrow Cu(acac) + H(acac)$.

9 Factors that Influence the Interaction Between the Metal Complex and the Oxide Surface

The interaction between the metal complex and the surface of the support is influenced by a number of factors. In the following, we list each factor and then attempt to rank them according to relative importance.

9.1 Stability of Metal Complex. – The stability of the metal complex towards ligand exchange reaction plays a major role in determining the mechanism by which the metal complex becomes anchored to the surface.[37] The ligand exchange mechanism for anchoring the metal complex to the surface proceeds as shown by the following scheme:

$$ML_n + H\text{-}OSu \Leftrightarrow ML_{n-1}OSu + HL$$

Here, we illustrate the exchange of an anionic ligand, L, for a surface oxide, ^{-}OSu, to yield the anchored metal complex ($ML_{n-1}OSu$) and the stable molecule HL. The stability of the metal complex for this exchange reaction determines the extent, if any, that the metal complex will surrender ligands. Thus, in the design of metal complexes to be precursors for the supported metal oxide, the catalytic scientist must consider the relative stability of the metal complex, ML_n, *versus* the stability of the partially-exchanged complex: $ML_{n-1}OSu$. While, data are not readily available for the stability of the surface complex ($ML_{n-1}OSu$) one could estimate the likelihood of the exchange reaction using -OH as a model for -OSu.

For these exchange reactions to occur at mild conditions using a liquid solvent, the geometry of the metal complex must allow the surface protons to gain access to the anionic ligand. Metal complexes showing square planar geometries permit access to the ligands and to the metal cation; whereas the octahedral geometry often frustrates the access to the ligands by surface protons. For example, the *tris*-$Cr(acac)_3$ complexes will not engage in ligand exchange reactions at low temperatures (< 100°C) as might be used when a liquid solvent is used. However, this problem can be circumvented when the catalyst preparation is accomplished at higher temperatures (>200°C) by sublimation of the volatile metal acetylacetonate in a gaseous solvent. At these higher temperatures, the kinetic energy of the system may be sufficiently high to overcome the energy barrier offered by the steric hindrance of the (acac) ligands in the $\mathbf{O_h}$ symmetry.

9.2 Surface Acidity/Basicity. – Van Veen[38, 39] and co-workers were one of the first groups to discuss the importance of surface acidity/basicity as it relates to the interactions that favor decorating metal oxide surfaces with certain metal complexes. We alluded to the acidity/basicity of the surface (*vide supra*) in describing how a metal complex may become subject to a reaction with the surface. Van Veen, *et al.* were interested in decorating surfaces with a wide range of metal acetylacetonate complexes to include Pt and Pd precursors. They showed that metal complexes that were susceptible to attack by acids were likely to decorate the surface of an acidic surface oxide by loss of a ligand to form a firm attachment with the surface.

9.3 Choice of Solvent. – In choosing a suitable solvent, the solubility of the precursor in the solvent is only one factor that influences the interaction of metal complexes and the surface of the support. The other factors that will influence this interaction are as follows:

1. the solvent must not compete with the precursor for sites on the surface,
2. the solvent must not coordinate to the precursor, and
3. the solvent must not react with the precursor or cause the precursor to decompose.

While water is used as a solvent in a majority of commercial adsorbent and catalyst preparations, we can find instances, where it violates all of the rules listed above. Many metal oxide supports have a high affinity for water, and in fact, supports such as silica, alumina and molecular sieves are used as drying agents for light gases. Water dissociates into protons and hydroxyl ions to a small extent ($K_w = 10^{-14}$), but when transition metal ions are introduced into water, it coordinates to many of these metal ions forming hydrates, and/or it reacts with the metal ions to form metal hydroxides. It is for these reasons, that we actively searched for an alternative method to prepare supported, mixed metal oxides using solvents such as acetonitrile that obey the three rules listed above for many systems including the perchlorate salts of polynuclear metal complexes.

Alcohols[17], such as methanol and ethanol, has been used as solvents for the metal acetylacetonate systems, usually under reflux conditions. For many of these metal acetylacetonates, these alcohols are good solvents and show only a low affinity for most supports. However, as we discussed earlier, some metal acetylacetonates form μ-alkoxy dimers when catalytic amounts of ^{-}OH are present. The unexpected formation of these dimeric precursors may frustrate attempts to form a supported metal oxide catalysts with isolated metal cations decorating the surface.

Aromatics[18] such as benzene and toluene have also been used as a solvent for highly chelated metal complexes, such as in the preparation of supported metal acetylacetonates. Aromatics are a good choice for metal complexes formed with aromatic ligands such as bipyridil. Tetrahydrofuran (THF) has also been used as a solvent to prepare silica-supported Pd that was generated by the decoration of the surface with $Pd(acac)_2$.[40]

9.4 Ionic Metal Complexes. – For supports such as silica and some zeolites, cation exchange is a viable mechanism to anchor the precursor; therefore, it would be advantageous to synthesize the precursor as a cation metal complex. While this consideration is obvious, this approach is frustrated because many simple metal complexes of catalytic significance are anions, such molybdates, vanadates, titanates, *etc.* Thus, one may design the appropriate metal complex using ligands such as alkoxyamines,[46] ethylenediamine,[41,42] diaminopropoxides,[47] acetylacetonates,[43,44,45] *etc.* often as the cationic salts of perchlorates or phosphorous hexafluorides.

For silica, the protons on the surface silanols are only weakly acidic, therefore ion exchange of these protons for a cation complex must be assisted by the use of a Lewis base to activate the surface protons as:

$$\text{SuO-H} + \text{base} \Leftrightarrow \text{SuO}^- + \text{H-base}$$

The base can be a co-solvent, such as trialkylamine, or the Lewis base function can be designed into the metal complex using O-bearing ligands such as alkoxyamines, acetylacetonates, *etc.* The optimum amount of co-solvent is 1-2 equivalents of the metal complex present in the solvent.[47] The nitrogen atoms in ethylenediamine are not sufficiently basic to activate the surface protons from silica, but ethylenediamine complexes of some metal cations, such as Cu(II), readily exchange for the protons in zeolites which are more acidic than silanol protons.[41]

9.5 Hydrogen Bonding. – The attachment of a metal complex to a surface by hydrogen bonding interactions has been observed when an organic solvent is used that is aprotic and non-coordinating. The nature of the hydrogen-bonding interactions can arise from surface OH groups and lone pair(s) of electrons on the oxygen or nitrogen atoms in the ligands of the complex (*e. g.*, ethanol amine, alkoxy amine, acetylacetonate, bipyridine). This means of attraction can be quite important when the ligands are able to fully engage the surface protons such as in the square planar, *bis* (acac) complexes.

Ranking the relative importance of these factors is difficult when one considers the proper metrics to use. That is, one could choose the initial loading of the metal complex on the solid after exposure to a solution containing the metal complex. Or, one could consider the retention of the metal complex to the surface when the solid has been washed repeatedly with fresh solvent before the system is heated to decompose the ligands. Another metric, is the dispersion of the supported metal ions after heating to remove the ligands. Still another metric is the variation in the dispersion when the solid has been heated to high temperatures (500°C) in oxidizing or reducing conditions. Consider the following examples from our own experience which illustrate how difficult the problem is for ranking the importance of these different factors.

For convenience, we use silica as a common support and acetonitrile as the common solvent then we rank the importance of the factors described as judged by the first metric: initial loading of complex on the solid. We observed that some charged species were retained on the surface at higher loadings than uncharged species using only hydrogen bonding interactions provided that the surface protons could be activated by either a co-solvent (triethyl amine) or by basic oxygen groups on the ligands (alkoxy amines). However, when no basic oxygen atoms were present in the ligands and no co-solvent was used, even strongly charged metal complexes, such as $Cu(en)_2^{2+}$, were only weakly attracted to the surface at low loadings. For this last case, the uncharged $Cu(acac)_2$ complex was retained to the surface by hydrogen bonding in higher loadings than the $Cu(en)_2^{2+}$. Next, consider the relative importance of these factors when the metric is the retention of the complex on the surface when fresh solvent is used. The uncharged metal complexes that don't experience ligand exchange with the surface were removed over time with washing whereas the charged species were retained at the monolayer loadings in the face of extensive washing with fresh solvent.

If we now turn to the stability of the uncharged metal complex towards ligand exchange, we find that this factor is of great importance in the initial loading of the metal ions on the surface. Consider the relative stabilities of the following metal acetylacetonates towards ligand exchange reactions: $Pt^{2+} \sim Pd^{2+} > Cr^{3+} > Cu^{2+} > Fe^{3+} > Co^{2+} \sim Mn^{2+}$. The first three metal acetylacetonates are very stable and do not easily loose ligands. These three (Pt^{2+}, Pd^{2+}, & Cr^{3+}) also don't decorate surfaces from liquid solutions with high loadings, *i. e.* > 1 wt% metal ion on support. The Cu^{2+} complex easily forms a monolayer without loosing its ligands; whereas Fe^{3+}, Co^{3+}, and Mn^{2+} all surrender some or all ligands while decorating the surfaces at high loadings. Here, we see that among the uncharged metal complexes, lability of the metal complex to ligand exchange plays an important role in determining the initial loading of the complex on the surface of the silica support.

For complexes that only hydrogen bond to the surface without ligand exchange, the symmetry of the complex plays an important role in decorating the surface with high loadings from liquid phase solutions at reflux and containing the metal complex. Consider the *tris*-$Cr^{3+}(acac)_3$ and the *bis*-$Cu^{2+}(acac)_2$ in acetonitrile over silica. The $\mathbf{O_h}$ symmetry of the *tris* complex precludes it forming high loadings of the complex when applied to the solid from a liquid solvent whereas the favorable $\mathbf{T_d}$ symmetry of the *bis* complex allows it to engage the surface through hydrogen bonding. However, this steric factor alone cannot explain the relatively low loadings observed for *bis*-$Pt(acac)_2$ or *bis*-$Pd(acac)_2$ on silica. We must now recognize that another factor is operative for these systems. These two, group VIII metal complexes do not readily accept ligands into the axial ligand positions whereas Cu^{2+} easily accepts an axial ligand. We observed the same result for the dihydrate of $Mg(acac)_2$ which easily surrendered one water molecule to form a monolayer on silica or on alumina.[42,43]

Having said this, we can put all of these considerations into the proper perspective by sharing how we would proceed with a catalyst preparation by an ideal approach. The support would be robust, would not be 'corroded' by the solvent, ligands, *etc*., and would show reactive hydroxyl groups whose site density can be adjusted by either thermal means (drying under vacuum) or by chemical means (reaction with strong base). Silica appears to fulfill this prescription for many systems. The solvent would dissolve the metal complex without reacting with or forming an adduct with either the metal complex or the support. We have used acetonitrile with excellent success. As a result of preliminary design of the catalytic ensemble, the ideal choice for a metal complex is one that is active for the desired catalysis when all of its ligands are intact. The metal complex would be a cation with stable ligands that contain strong Lewis bases, preferable in the form of oxygen atoms and these ligands do not react with the solvent or other metal complexes. The anion to the metal complex is perchlorate. The metal complexes then interact with the support through the Lewis base functional groups in the stable ligands by first activating the surface silanols to form surface siloxides. The perchlorate anions form volatile adducts with the surface protons (*e. g*., perchloric acid hydrates that sublime at 90°C). The metal complex cations are retained on the surface by ionic bonding interactions and self-assemble into a close-packed monolayer. This ideal catalyst synthesis was realized by our first effort and the properties were determined by a simple semi-empirical quantum mechanics calculation before we measured its chemical properties.[7] The Extended Huckel MO program accurately predicted the effects for changing M^{3+} upon the Bronsted-catalyzed reactivity, the enthalpy of adsorption for ammonia, the vibrational frequencies of the bridging OH groups, and the dipole moment of the O–H bond in the complex.[7]

10 The Chemistry of Organic Supports

The previous discussions centered on the decorating of oxide surfaces with metal complexes and as we have seen, this decorating methodology exploits the reaction of surface OH groups with the Lewis acid or Lewis base functionalities in the metal complexes and/or depends upon hydrogen bonding interactions between the surface and the metal complex. This fundamental chemistry can be used to decorate organic supports provided that surface OH groups can be developed on this support. Alternatively, one could functionalize the surface of the organic support with an organic ligand that could then interact with the target metal complex by a favorable ligand exchange reaction. In the following we show one example for which metal complexes have been grafted onto organic supports.

10.1 $M^{2+}(acac)_2$, (M = Cu, Co)/C. – Valente, *et al.*, recently described their attempts to affix Cu^{2+} or Co^{2+} to the surface of an activated carbon that had been functionalized with hexanediamine using thionyl chloride.[44] Preparation of the carbon included 1) oxidation of the surface with nitric acid following by washing to neutral pH and drying at 90°C, 2) refluxing this oxidized sample in thionyl chloride (SO_2Cl_2) for 1 hour followed by distilling the SO_2Cl_2 from the sample, 3) this sample was then contacted with hexanediamine for 6 hours, followed by washing with ethanol and drying at 80°C for 2 hours, and 4) reacted with a solution of $M^{2+}(acac)_2$ (M = Cu or Co) in chloroform at reflux for 16 hours. This material was washed with ethanol and dried at 80°C.

The nitric acid wash converts the carbon surface to one that is rich in carboxylic acid groups. These -COOH groups react readily with thionyl chloride to form the corresponding surface acyl chloride groups -COCl. This surface acyl chloride can then react with one of the amine groups in the hexanediamine so as to anchor the molecule to the surface. The resulting tether then shows a primary amine that will presumably react with the $M^{2+}(acac)_2$ (M = Cu or Co) so as to anchor the metal complex to the surface. The entire scheme is shown below (Scheme I).

The metal complexes, $M^{2+}(acac)_2$ (M = Cu or Co), were observed to be bonded to the surface tether by a Schiff condensation reaction with the primary amine. The metal cations in these tethered complexes were used to activate *t*-butylhydroperoxide into radicals that reacted with the substrate, *cis*-pinane, to give mainly 2-pinane hydroperoxide with no observed formation of 2-pinanol. The selectivity to the pinane hydroperoxide was 93% for the immobilized Co(II) and 84% for the immobilized Cu(II) at 91% substrate conversion. Blank reaction tests using the (1) oxidized carbon and (2) oxidized and hxd-functionalized carbon showed much lower activities (~10-15% conversion) thus indicating that the tethered metal complexes were largely responsible for the observed activity. The subsequent reuse of the tethered catalysts eventually showed a constant activity (~ 60% conversion) with reuse suggesting that the tethering method was effective in firmly attaching the metal ion to the surface.

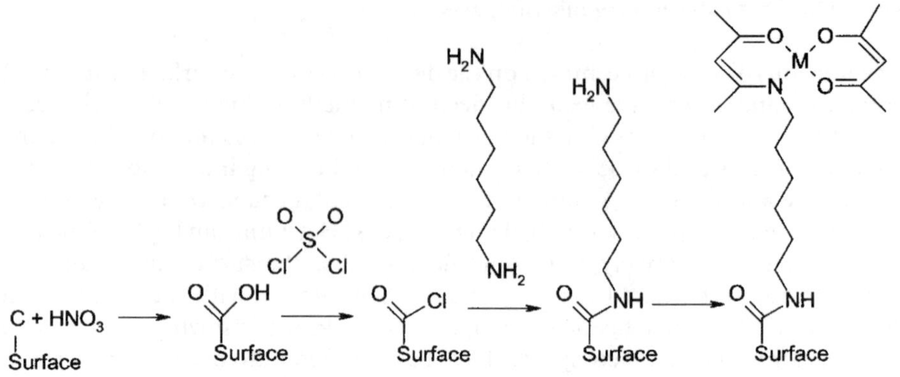

Scheme I

11 Characterization of Supported Metal Complexes

In the following we list common methods that have been used to characterize supported metal complexes and show how these techniques can be used to learn more about the surface-metal complex morphologies and chemical properties.

11.1 Chemical Analysis of Supported Neutral Metal Complexes. – Certain aspects of the gross morphology of supported metal complexes can be inferred from a consideration of the chemical analyses of the system. For example, one can determine if the complex has reacted with the support by comparing the ratio of certain elements present in the ligands to the amount of metal in the system. In the case of metal acetylacetonates, one often uses elemental analyses of carbon and the metal to calculate the ratio of moles (acac) ligands to the mole of metal ion. This value is referred to as the R-value by some authors.[45] For metal acetylacetonates that are stable in weak acids and bases, such as $Cu(acac)_2$, the R-value is 2 for both supported and unsupported metal complex. *tris*-Iron(III) acetylacetonate is an example of a metal complex that reacts with a support, silica, at room temperature upon contacting silica from solution of acetonitrile by loosing an (acac) ligand so that the R-value for the supported metal complex is less than that for the unsupported metal complex. The determination of the R-value is complicated when the support also contains significant amount of carbon, as in a surface carbonate. The use of hydrogen analyses to determine the R-value is frustrated by the significant amount of hydrogen that is always present in supports either in the form of surface -OH groups or in the form of molecular water.

11.2 Chemical Analysis of Supported Ionic Metal Complexes. – When the metal complex forms an ionic solid that can participate by complete ion exchange of the anion for surface anionic species, it may be possible to document the completion of monolayer formation by quantitative analyses for the counter ion to the metal complex. Our first studies demonstrated the usefulness of this approach as all of these metal complexes were synthesized as the perchlorate salts and the metal complexes surrendered the perchlorate anions in exchange for siloxides.[19] The perchlorate anions combined with surface protons and molecular water to form volatile hydrates of perchloric acid upon heating to 90°C. Thus, the cationic complexes formed a monolayer film without retaining perchlorate anions; however, subsequent layers of the cation formed with the incorporation of perchlorate anions. Data of perchlorate analyses at loadings in excess of monolayer formation could be used to define with precision the monolayer loading of complex. It was determined that the complexes forming the monolayer were bound tightly to the surface and thus, samples having multiple layers of the complex could be washed with the solvent to obtain a monolayer sample.

We illustrate this idea further using the perchlorate analysis to determine the completion of a monolayer of the supported metal complexes (Fig. 9).[46, 47] Here we show the perchlorate analysis of silica-supported samples prepared from the perchlorate salts of three different copper metal complexes: 1) Cu^{2+} $(C_2H_8N_2)(ClO_4)_2$ (ethylenediamine), 2) Cu^{2+} $(C_8H_{21}N_2O_4)$ (ClO_4) (DETA), which is prepared from dihydroxydiethylamine ligands and 3) the complex $\{[O\text{-}CH_2CH_2\text{-}N\text{-}R_2Cu\text{-}\mu\text{-}OH]_6M\}^{3+}$ $(ClO_4)_3$ $(Cu_6(OH)_6M)$ prepared from N-N-diethylethanolamine ligands. The slopes of these curves reflect the perchlorate

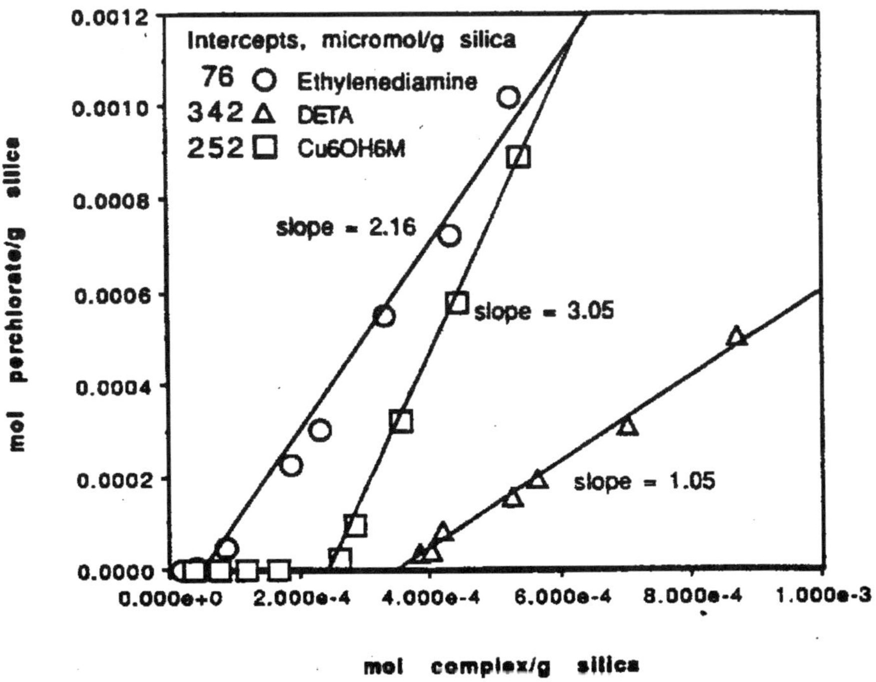

Figure 9 *Perchlorate analysis of silica supported metal complexes.*

incorporation in the layers of metal complex in excess of a monolayer; whereas, the intercepts represent the loading of metal complexes that formed a monolayer. The copper ethylenediamine complex uses 2 perchlorate anions to balance the charge of the Cu(II), whereas the incorporation of perchlorate anion is 1 per complex in the copper(II) dihydroxydiethylamine complex. The hexameric copper μ-OH-M(III) complex uses three perchlorate anions to balance the overall charge of 3+ of the cation. The intercept data can be compared to the theoretical monolayer loading determined from a consideration of the sizes of the metal complex 'footprint' on the surface and the surface area of the silica, which we found to be 200 m^2/g by BET surface area analysis.[19, 47] The projected sizes of these complexes were as follows: copper ethylenediamine = 52.90 $Å^2$/molecule; *bis*-copper-2,2′- dihydroxydiethylamine = 59.26- 66.99 $Å^2$/molecule and hexameric copper μ-OH M(III) = 132.8 $Å^2$/molecule using molecular models of the complexes for which the van der Waals radii were used to compute the sizes. From these values we estimated the theoretical loadings of a monolayer to be 628, 463-518 and 250 μmole/g-silica, respectively. Whereas these predictions agreed closely with the observed monolayer loadings for two of the metal complexes (489 vs. 463-518; and 252 vs. 250 μmole/g-silica), the observed monolayer loading was much lower for the copper ethylenediamine complex: 76 *versus* 628 μmole/g silica.

This result can be understood when we consider the Lewis basicity of the ligands in each of the metal complexes. The nitrogen's in the copper ethylenediamine metal complex show all of the electrons in the frontier orbitals (HOMO/LUMO) to be fully involved in bonding and thus cannot function as a Lewis base whereas, the oxygen atoms in the diethoxyamine and the N-N-diethylethanolamine ligands do show lone pairs that can be considered Lewis bases. It is the lone pairs of the strong Lewis bases that activate the protons from the surface OH groups and thus permit ion exchange to occur. On the other hand, the ethylenediamine ligands are much weaker bases and cannot activate the surface protons and thus very little ion exchange is possible. Venable confirmed this hypothesis when she showed that much higher loadings of copper ethylenediamine could be obtained on silica when a co-solvent, triethylamine, was used with acetonitrile.[16] Choksi confirmed the results of Venable by reporting Cu metal loadings up to 2.4 wt% in the $Cu(en)_2^{2+}$ and extended it to include the following metal cation complexes: Cu-bipy monomer, and Cu-bipy dimer, Cu-dapo.[48] This suite of copper metal complexes was chosen to illustrate the effect of the basicity of the functional groups in the ligands upon the binding of the metal cation complexes to the silica. The (en) and (bipy) ligands showed nitrogen atoms that are weak Lewis bases; however, the (dapo) ligands show oxygen atoms that are strong Lewis bases. Mild conditions were chosen and acetonitrile was the solvent so that the complexes remained intact during the contacting of the solid with the solution containing the metal complexes. Without the triethylamine co-solvent, the Cu-en, Cu-bipy monomer, Cu-bipy dimer samples gave very low loadings (<1.0 wt %) of metal complex/silica and much higher loadings were realized with the addition of the co-solvent (>3.0 wt %). On the other hand, the Cu-dapo showed high loadings without the co-solvent.

Thus, we suggest that the ion exchange mechanism can be fostered when the ligands within the metal complex show strong Lewis base functions. As a result, we try to use these type of ligands when designing a metal complex to be used as the source of the neutral ion in a supported catalyst.

One last comment is appropriate regarding the bonding of the *bis*-copper-2,2′-dihydroxydiethylamine to the silica surface. The ligands are not equivalent in that one ligand shows an extra hydrogen; thus, the reason for the odd number of hydrogen's in the parent complex. Therefore, the hydrogen-deficient HDETA, anionic ligand reacts with a proton on the silica surface upon bonding to form the stable compound: diethanolamine, H_2DETA ($C_4H_{12}O_2N$). This reaction permits the surface siloxide to bond to the copper(II) and thus anchor the complex firmly on the surface. A rich knowledge of ligand-metal chemistry is helpful to fully exploit these ideas in decorating surface oxides with metal complexes.

The results of Cu loadings observed for samples prepared by ion exchange of silica with $[Cu(en)_2]^{2+}$ appear to be confirmed by the recent data reported by Toupance, *et al* on their attempts to prepared Cu/silica solids.[49] They reported Cu loadings < 2 wt% when the cation was the copper ethylenediamine; however, they did not mention the use of a co-solvent.

11.3 Thermal Analysis. – Several researchers have used thermal analysis as a means to study the structure and chemistry of supported metal complexes.[50] This technique can be quite informative provided that the support oxide does not contribute much to the mass changes of the sample upon heating. Silica is a stable support that shows only small relative mass changes at temperatures lower than 600°C. Supports that form carbonates are less appealing as this mass change must be considered when the carbon-bearing parts of the metal complexes are decomposed during thermolysis.

Mass changes that attend the loss of ligands upon heating in inert atmospheres are often used to corroborate the elemental analyses data of the supported metal complexes to define the stoichiometry of the complex before thermolysis. When the ligands are anionic and when these ligands react with the surface protons to form stable molecules during the thermal treatment, thermal analysis can provide useful information about the structure of the supported metal complex provided that the evolved stable molecule (*e. g.*, pentanedione) does not react with metal ions in the support (*e. g.*, alumina). For example, elemental analyses of solids formed by the reaction of $Cu^{2+}(acac)_2$ and silica suggested that the surface supported metal complex formed a monolayer with both ligands intact.[11,27] Mass changes attending the thermolysis of these samples were consistent with a model of the supported complex having both ligands present before the samples were heated. Moreover, two regions of mass loss were observed ($\sim$250 and $\sim$290°C) which suggested that the ligands decomposed at different temperatures by different mechanisms.[11,27] We speculated that the lower temperature decomposition occurred as a result of the reaction of surface proton with one acetylacetonate anion and the higher temperature decomposition was 'uncatalyzed'. This hypothesis appeared to be of a general nature as it was confirmed subsequently by others who were investigating a different metal acetylacetonate: supported

$Ru(acac)_3$.[51] We show here the thermal analysis (TA) data presented by these researchers that illustrate well the decomposition of the ligands at the two different temperatures: 210 & 270°C (Figs. 10-12).

This result is similar to what we reported earlier for the $Cu(acac)_2$/silica system where the temperatures for decomposition were 250 and 290°C. These figures show the mass change (TG), differential mass change (DTG) and heat flow (DTA) for samples of $Ru(acac)_3$ on silica at three different loadings 0.2, 1.0 and 5.0 wt% loadings. Notice how the mass change and the differential mass change data scale with the loadings of $Ru(acac)_3$ as it decreased from 5.0 to 0.2 wt% (Figures 10 – 12) and how the differential heat flow rates decrease with decreasing weight loading of ligand on silica.

Also, notice how the exotherms for the heat flow data occur at the same

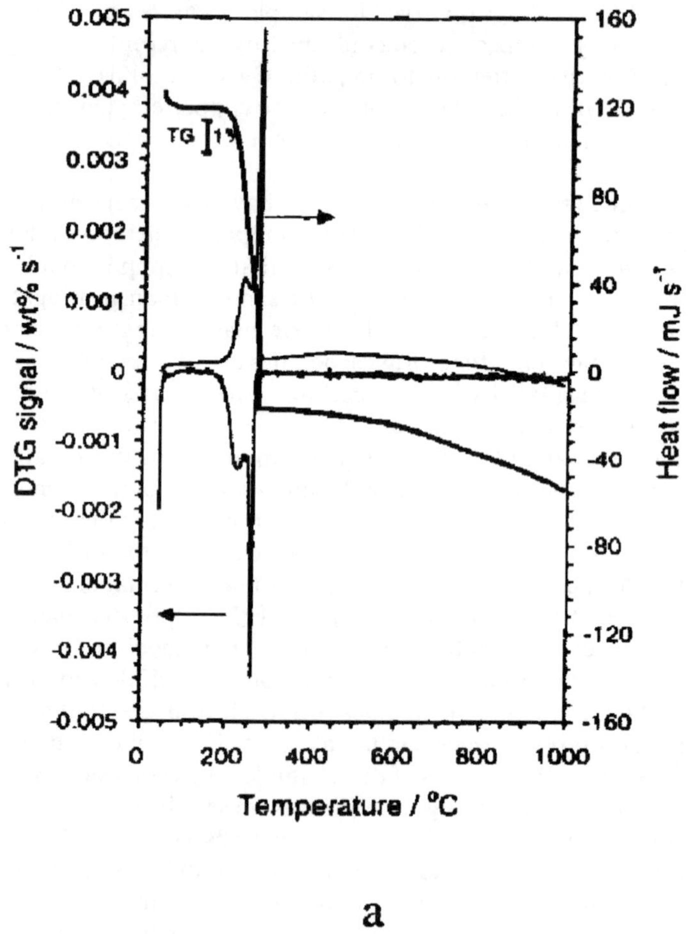

Figure 10 *TA data for 5 wt% $Ru(acac)_3$/silica*

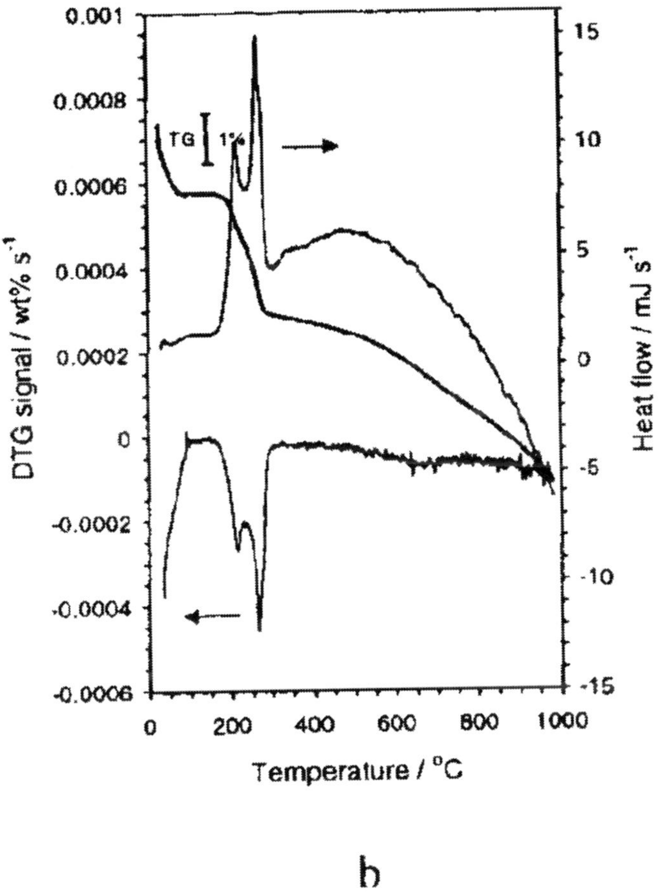

Figure 11 *TA data for 1.0 wt% Ru(acac)$_3$/silica*

temperatures for the minima for the DTG data suggesting that the ligands decompose by an exothermic reaction. Finally, notice that the extent of mass change occurring at the lower temperature is about 2/3[RDS] of that occurring at the higher temperature, as shown in Fig. 11. These authors discussed these results in the context of models where the Ru(acac)$_3$ metal complex interacted with two surface protons residing on adjacent silanol groups (scheme II).

Notice here that only two of the three (acac) ligands can interact with the surface protons, thus it appears that only 2/3[RDS] of the ligands can be lost by proton-assisted thermolysis. One of the samples, 1 wt% Ru(acac)$_3$ appears to show a weight loss in the TG spectrum consistent with this prediction. It must be remarked that the data of this same sample agreed in the predicted and observed weight loss upon thermolysis (2.6%); whereas the other two samples showed different experimental and predicted weight losses.

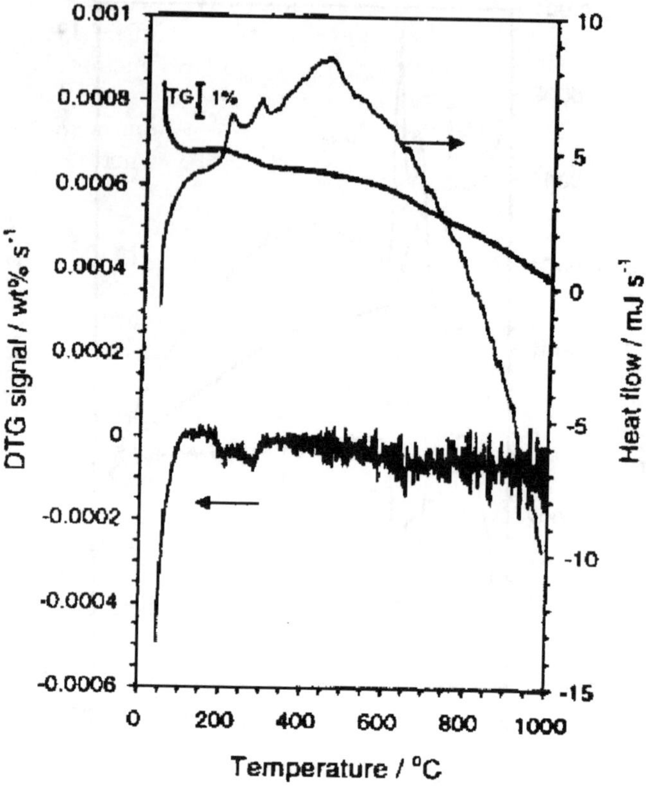

C

Figure 12 *TA data for 0.2 wt% Ru(acac)$_3$/silica*

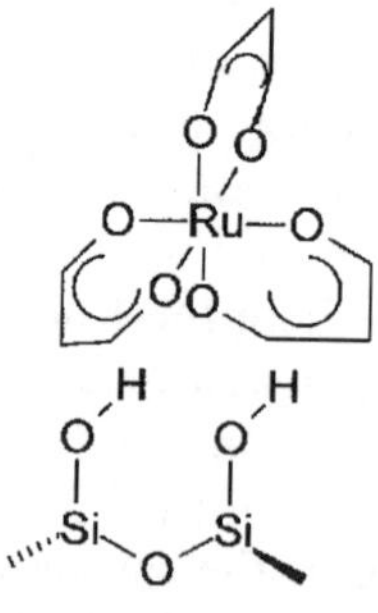

Scheme II

11.4 Spectroscopy. – In the following paragraphs we discuss how various spectroscopy tools have been used to illuminate the rich surface chemistry of supported polynuclear metal complexes. These methods include infrared (both diffuse reflectance and transmission), ultraviolet and visible (UV-Vis), x-ray photoelectron spectroscopy (XPS), and x-ray absorption, near edge spectroscopy's (XANES) and extended X-ray absorption fine structure (EXAFS) spectroscopy's.

11.4.1 InfraRed Spectroscopy. – Infrared spectroscopy (IR) has been used for over 50 years[52] to examine solid catalysts and the events that accompany their use. Since that time, a large number of publications have chronicled the strengths and weakness for using IR as a means to characterize the surface of solids. Well-defined, metal complexes supported on certain solids are amenable to careful characterizations by IR. For one application of this technique, qualitative aspects of the metal complex structure can be deduced as a function of metal loading on the support, or with changes in the pretreatment, *etc.*

11.4.1.1 Cu(acac)$_2$/silica. For some metal complexes, the morphology of the supported metal complexes can be inferred from a detailed investigation of the IR spectrum. Copper acetylacetonate is one such complex for it remains intact when it binds to silica from liquid solution at room temperature. Moreover, the double overtone vibration of the ring carbon for the deformation vibration γ(C-H) shifts up field by 20 cm^{-1} when it forms a monolayer or less on the silica. Thus, one can devise a protocol to examine in a systematic manner the IR spectrum of a series of samples having a fractional coverage near unity so as to determine the loading at saturation and compare with the predictions of a molecular model. We were able to make such a determination in a model system and found good agreement between predicted and observed loadings for monolayer saturation of an amorphous, non-porous silica (Cab-O-Sil).

Mitchell, *et al.* described the diffuse reflectance IR spectra of silica-supported Cu(acac)$_2$ forming prepared by extended refluxing of the acetonitrile solvent containing silica and the metal complexes.[37] They attempted to prepare samples showing both monolayers and multiple layers of Cu(acac)$_2$ on the surface of this solid and showed that for metal complex loadings less than a monolayer (Fig. 13, 1.42 x 10^{14} molecules/cm^2) the DRIFT spectra were missing one vibration that was present in the samples containing more than a monolayer and also present in the spectrum of the polycrystalline metal complex.

They considered not only these qualitative aspects of the IR spectrum but also the quantitative changes in the absorbance with loading of metal complex to conclude that 1) the disappearance of the 1552 cm^{-1} peak is related to the mechanism that causes the disappearance of the 785 cm^{-1} fundamental vibration. The adsorbed complexes in intimate contact with the surface contribute no absorption peaks of the complex in the spectral region below 1000 cm^{-1}. Thus, the disappearance of the 785 and 1552 cm^{-1} vibrations are characteristic of complexes in close contact with the surface.

Others have used IR to describe the effect of washing and drying protocols on

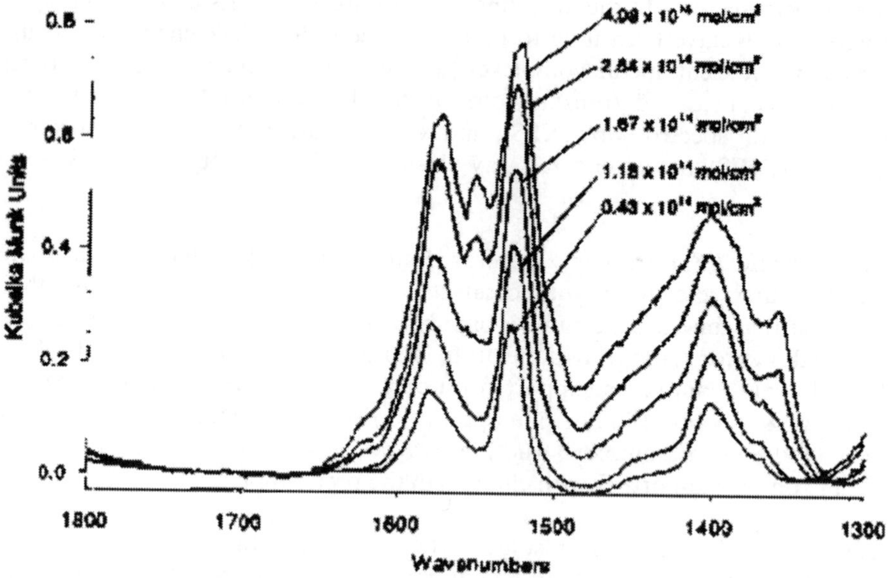

Figure 13 *DRIFTS of supported* $Cu(acac)_2$ *on silica at loadings of 0.43, 1.18, 1.47, 2.64, and* 4.08×10^{14} *molecules/cm*^{2}

the state of the dispersion of the metal complexes on the surface. Baltes, *et al.* (Fig. 14) described their attempts to alter the interactions between silica and $Cu(acac)_2$ when the solvent was toluene.[45] They reported that a sample initially showing some metal complexes not in contact with the surface can be treated by repeated washing with fresh solvent to produce a sample in which most of the metal complexes were distributed as a single layer on the support. Drying of the sample under a vacuum improved the dispersion of the metal complexes in the sample. Thus, the decoration of the surface metal complexes can be improved by the technique of repeated washing in the toluene solvent.

The thermal decomposition of the silica supported $Cu(acac)_2$ could be followed by transmission IR of a self-supporting wafer of the supported metal complex in a controlled atmosphere cell that can be heated to 400°C.[11] When the supported metal complex at a monolayer loading (~3.5 wt% Cu) was heated to 250°C, the IR spectrum changed to show the reappearance of the vibration at 1552 cm^{-1}, which is characteristic of the double overtone of the 2γ(C-H). TGA data showed that these conditions were sufficient to just remove only one (acac) ligand, probably by proton-assisted thermolysis.[11] The appearance of the vibration at 1552 cm^{-1} suggested that this remaining (acac) ligand was not interacting with the surface (Fig. 15) and therefore would require a higher temperature to remove it from the complex. Heating the sample to 350°C did remove all traces of the ligand in the IR pattern.

11.4.1.2 $Ru(acac)_3/SiO_2$. The differences discussed here for the IR spectra of the supported and supported copper acetylacetonate complex are not observed in

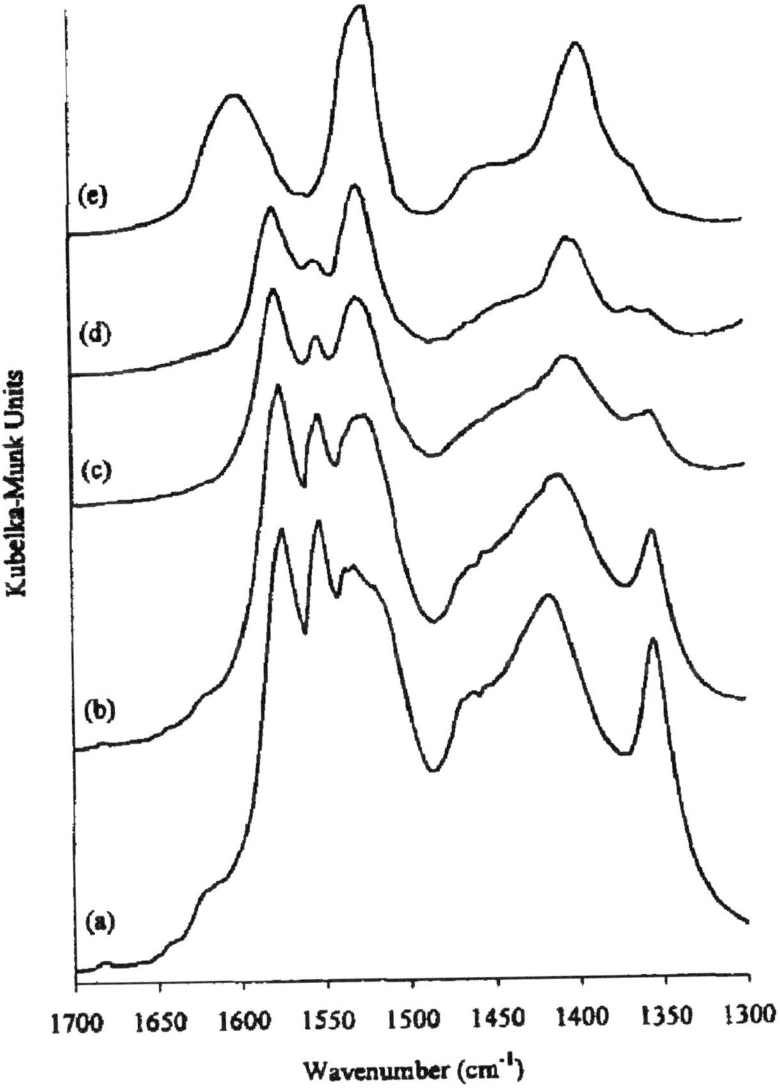

Figure 14 *DRIFTS of a) pure Cu(acac)$_2$; b) silica sample after liquid-phase modification with Cu(acac)$_2$; c) sample b after 7 washing cycles; d) silica sample after liquid-phase modification, 7 washing cycles, and drying in a vacuum at 80°C; e) alumina sample after liquid-phase modification*

the IR spectra for many metal complexes. For example, van Veen *et al*,[53, 54, 55] showed that the IR spectra does not change when Ru(acac)$_3$ is supported on either silica or alumina. Thus, the use of spectroscopy to interrogate for monolayer completion is not a general tool. However, this tool can often be used to document the interaction between ligands and structures on the surface of the oxide and to show when ligands have reacted with metal ions of the support. This

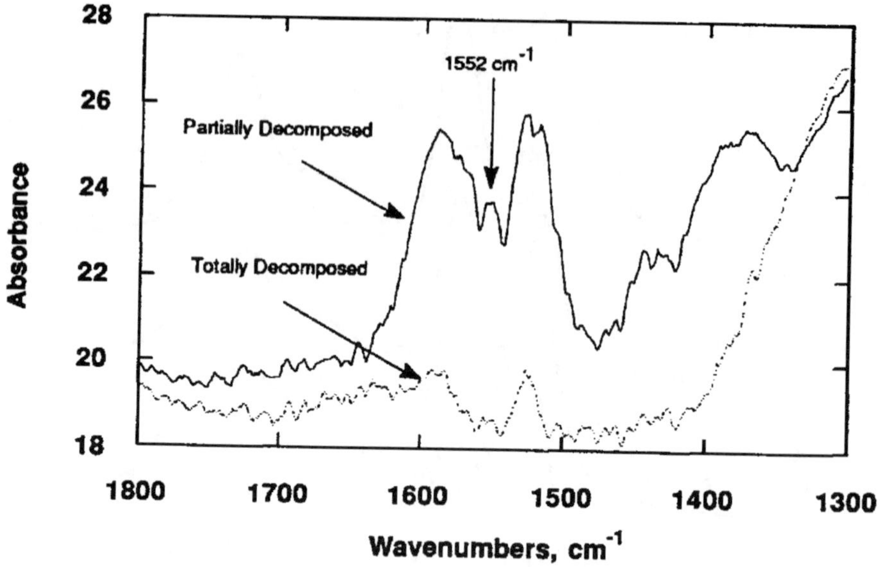

Figure 15 *Transmission IR of Partially-Decomposed Cu(acac)$_2$/silica*

last phenomenon has been reported often when metal acetylacetonates are supported on alumina to show the appearance of the Al(acac)$_x$ surface species.[56]

11.4.1.3 Fe(acac)$_x$/ZrO$_2$. Van Der Voort, *et al.* showed an interesting application of this technique to describe how Fe metal complexes decorate the surface of zirconia from liquid solutions of Fe(acac)$_3$ dissolved in dry toluene.[57] They show that Fe(acac)$_3$ readily reacts with the zirconia surface at room temperature from the liquid-phase and that Zr-(acac)$_x$ species are also formed presumably from the reaction of H(acac) with the zirconia surface (Fig 16).

The Fe(acac)$_{3-x}$ species are unstable on this surface so that the characteristic spectrum of iron acetylacetonate is not observed but instead a Fe-OH species is observed indicating that the iron has been grafted to the surface of the zirconia. The *tris*-Fe^{3+}(acac)$_3$ does not decompose completely on other surfaces, such as silica and titania[58], but rather it may loose 1-2 ligands upon contacting the surface. Thus, the intrinsic chemistry of the surface can have a significant effect upon the reactivity of the Fe(acac)$_3$ towards that surface. Upon heating to 300°C in air all evidence of the (acac) ligands had disappeared.

11.4.1.4 Pd(acac)$_2$/SiO$_2$. Daniell, *et al.*, prepared Pd/silica samples using Pd(acac)$_2$ dissolved in tetrahydrofuran under reflux to be subsequently examined by IR of the chemisorbed CO.[41] These researchers examined the influence of the pretreating atmospheres on the nature of the Pd surface (Pd oxidation state and size of Pd particles) created under the prescribed conditions and how this surface would chemisorb the CO probe molecule. The pretreatment atmospheres used for the decomposition of the supported Pd(acac)$_2$ were: O_2, H_2, CO, N_2, and

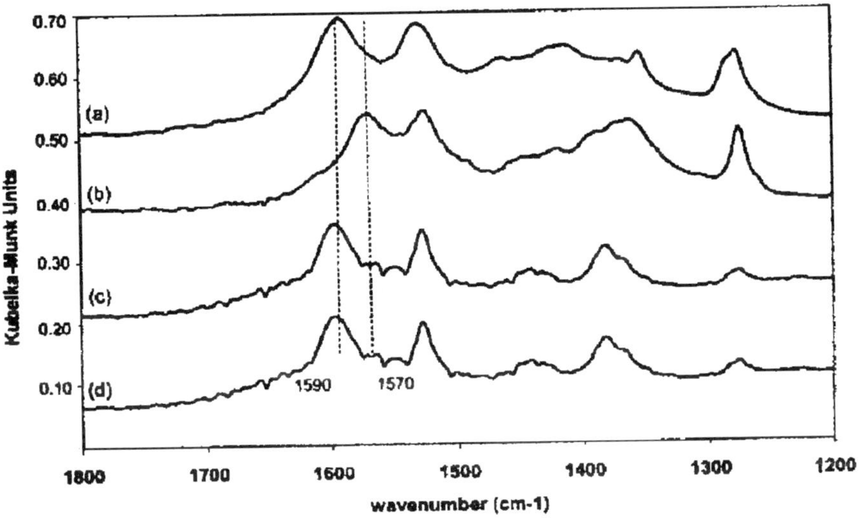

Figure 16 *Diffuse reflectance IR spectra of a) $Zr(acac)_4$; b) $Fe(acac)_3$; c) $ZrO_2/Fe(acac)_3$; and d) $ZrO_2/Hacac$*

vacuum at 570 K. Under mild conditions using CO as the reductant, Pd metal particles were formed, but when oxidizing conditions were established (O_2), particles containing Pd^{2+} were also formed. Carbide-like deposits were formed on the silica surface under strongly reducing conditions (H_2), and these deposits prevented access of the Pd metal sites for the adsorbate CO. Under neutral conditions (N_2), decomposition of the (acac) ligands led to partial reduction of the metal surface which showed several oxidation states.

11.4.2 UV-Vis. Spectroscopic measurements in the ultraviolet and visible range of the electronic spectrum (UV-Vis) can be used to probe electronic transitions in certain metal atoms and ion complexes. The energy of an electronic transition can depend upon the symmetry of the metal ion being different for transitions in a metal complex displaying tetrahedral (**T_d**) symmetry from the same metal showing an octahedral (**O_h**) symmetry. Thus, it is possible to use UV-Vis spectroscopy to interrogate the symmetry of certain metal ions bound to oxide surfaces. We show here a few examples of the use of UV-Vis spectroscopy to characterized supported metal oxides.

11.4.2.1 Supported VO_x/MCM-48. The charge transfer (CT) transitions between the oxygen anions and the central V cation result in strong absorptions of light in the UV-Vis region. Moreover, these electronic CT energies depend upon the size of the VO_x ensemble, being higher for isolated V cations as opposed to the lower energies observed for polymeric VO_x. In addition, the energy of the electronic CT also depends upon the local environment of the V centers whereby the energy for electronic CT in tetrahedral V (λ = 330 nm) is different from the energy of V in square pyramid (λ = 415 nm) and finally, the energy of the octahedrally coordinated V shows an absorption at $\lambda = 440$ nm. Morey, *et al.* exploited this science to characterize the V environments of VO_x supported on MCM-48.[59] Samples of MCM-48 were soaked in a dry hexane solution of vanadyl tri-*iso*propoxide [$(i\text{PrO})_3V{=}O$], filtered, dried, and calcined under dry O_2. These very dry conditions were necessary to synthesize the desired, dry product which appeared colorless in this state. It was anticipated that in this anhydrous state, the surface silanols would react with the metal complex to yield *iso*-propanol and a V=O grafted to the MCM surface by a SiO- to yield several moieties: 1) single Si-O-Vi=$O(\text{-OiPr})_2$ bridges, called A sites; 2) multiple $(\text{SiO-})_n\text{-V}{=}O(\text{-OiPr})_{3-n}$ called either B (n = 2) or C (n = 3) sites. These ester links will not survive in a moist environment thus UV-Vis characterizations were used to probe how the symmetry of the V cation center changes with exposure to water vapor.

The anhydrous VO_x/MCM-48 samples showed a broad peak centered at 325 nm (Figure 17) which was attributed to pseudo-tetrahedral $O_{3/2}\text{-V}{=}O$ center and a weak shoulder 415 nm which was assigned to the square pyramidal symmetry of 4 equatorial ligands and one axial oxygen. The spectra changed with time whereby the peak at 315 nm decreased, the peak at 415 nm first increased in intensity for an exposure times of 10 minutes (b) and then decreased at 30 minutes (c) with the appearance of a band at 440 nm which grew in size upon exposure to water vapor (ambient conditions, d). The peak at 440 nm was

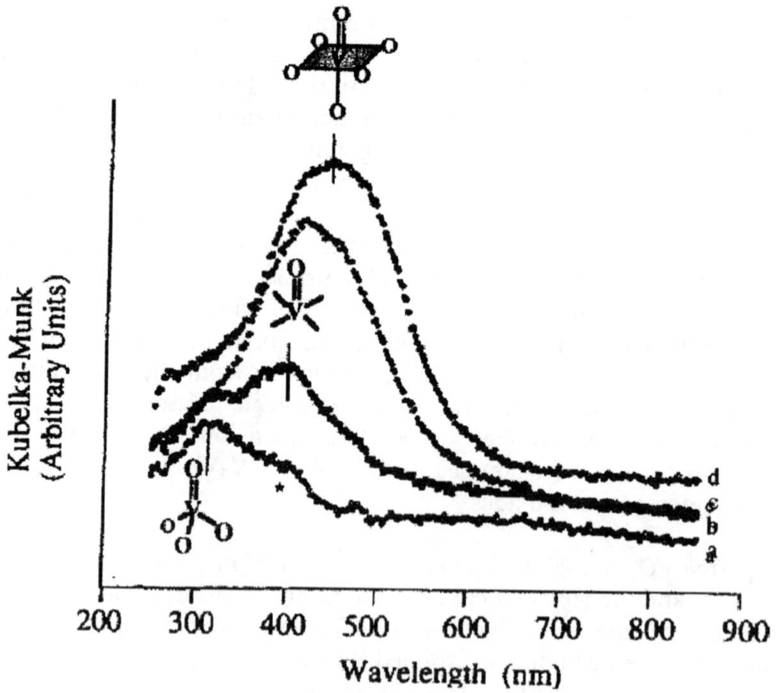

Figure 17 *UV-Vis VO_x/MCM-48*

attributed to octahedrally-coordinated V. These UV-Vis spectra were interpreted by a model in which water reacted with the surface species that was tetrahedrally-coordinated $O_{3/2}$-V=O [I] to break the three SiO-V bonds to form the square pyramid structure [II] which then reacted further to give the doubly-hydrated $\mathbf{O_h}$ coordinated V=O [III].

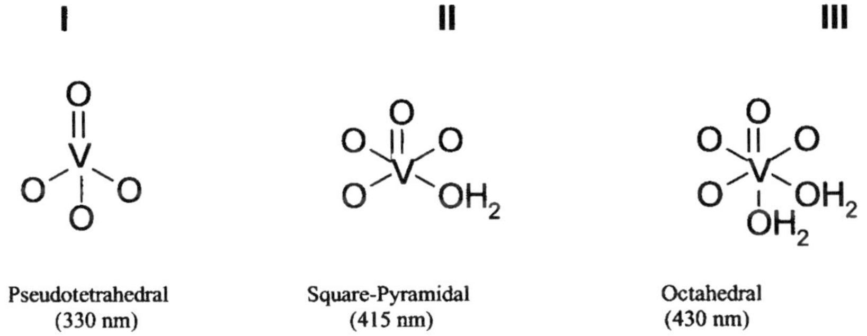

Scheme III

Subsequently, these samples were dehydrated and rehydrated for up to five cycles and the UV-Vis spectra recorded. This study showed that the phenomenon was reversible.

11.4.2.2 Supported CuO_x/silica. Toupance *et al.* reported the UV-Vis characterizations of silica-supported copper oxide.[48] Samples prepared by impregnation from an aqueous solution of copper nitrate to produce Cu loadings at 4 wt% and 16 wt% were first dried at 25°C (a); 100°C for 3 hours (b); 24 hours (c); and 100 hours (d) then these samples were calcined at 450°C (e); (f); (g); and (h), respectively. The 4 wt% Cu sample when dried showed a UV-Vis spectrum that was consistent with Cu^{2+} d-d transitions which shifted from a wavelength of 800 nm to 730 nm for increased temperature and duration of drying time. This shift to shorter wavelengths with increasing duration of drying was attributed to a change in the Cu environment to one in which some of the nitrate ligands were replaced with hydroxide ligands. The sample showing 16 wt% Cu and calcined at 450°C were also examined by UV-Vis to show large amounts of CuO as made evident by the sharp change in absorbance at 900 nm.

11.4.3 XPS. X-ray photoelectron spectroscopy (XPS) is a useful technique for probing the average oxidation state of the surface region to a depth ~ 10 nm. For supported metal oxides, this technique can yield information relating to the effect of metal ion dispersion and the effect of the support on stabilizing the oxidation state(s) of metal ions under reducing and oxidizing conditions.

Blackman, *et al.*[24] used the gas-phase technique, also known as atomic layer epitaxy (ALE), to develop controlled loadings of Co on silica using a $Co(acac)_3$ precursor (Fig. 18). By this technique they obtained loadings of Co from 5.7 to

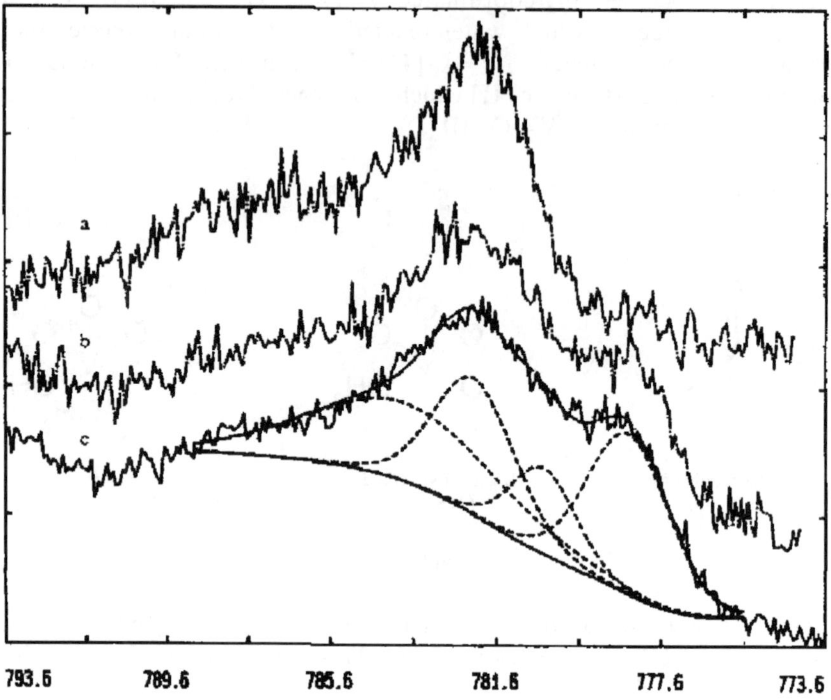

Figure 18 *XPS of supported Co samples prepared in either (a) 1, (b) 3, or (c) 5 preparation cycles*

19.5 wt% with % metal dispersion decreasing from 19 to 2.2%, respectively, with increasing metal loadings. These supported Co samples showed low degrees of reduction from 12 to 64%. XPS was used to interrogate the Co environment in these samples that were prepared in either 1, 3, or 5 preparation cycles having Co metal loadings of 5.7, 13.4, and 19 wt%. Deconvolution of the spectra (Fig. 18) suggested that least two, large peaks could be observed: 778.3 eV (Co^0) and 781.4 eV (Co^{2+}) and that two other small peaks could be present. Samples having the higher Co loadings (b and c) were easier to reduce than the sample at the lower metal loading (a). This result demonstrates how the successive applications of the metal complexes on the support can be used to produce layers of metal oxides having different properties. One can also imagine using this procedure to produce alternating thin films containing different metal oxides.

11.4.4 XANES. – $Au(CH_3)_2(acac)/MgO$[60] X-ray absorption, near edge spectroscopy (XANES) was used to determine the oxidation state of MgO-supported Au particles. These samples were prepared by decorating the surface of MgO with a Au(III) complex: $Au(CH_3)_2(acac)$. The samples were then analyzed for Au oxidation state as a function of the pretreatment gas (He or H_2) and pretreatment method. With increasing temperature of treatment, in either He or H_2, the Au was reduced from the cation to the metal. In the absence of H_2, it was speculated that the hydrocarbon fragments, either methyl or acetylacetonate, produced upon thermolysis the agency necessary for reduction of the metal ions.

11.4.5 EXAFS. $Au(CH_3)_2(acac)/MgO$.[59] X-ray absorption fine structure spectroscopy (EXAFS) was used to study the local structure of MgO-supported gold samples prepared by decorating the MgO with $Au(CH_3)_2(acac)$. These studies were completed for the sample without any post-synthesis treatment ('initial structure') and for samples developed upon heating this sample in either He or H_2 atmospheres.

The initial structure showed no significant Au-Au contribution which is consistent with mononuclear Au complexes that are isolated as a result of the effects of the Au atoms being fully coordinated by the ligands. Upon heating to temperatures where the methyl and acac ligands were removed by thermolysis, the Au-Au first shell coordination numbers increased from 1.1 to 9.4 suggesting that the average number of Au atoms/nanoparticle increased from 2 to ~80. For a sample treated at 373 K in He, the first shell Au-Au coordination number was 4+/0.4 with a bond distance of 2.82+/-0.02 A and a second-shell Au-Au coordination number of 1+/-0.1 at a distance of 4.02+/-0.04 A. One could model these results by a 6-atom octahedral cluster of gold. These nanoparticles still showed evidence of ligands chelating to the gold atoms as made evident by IR.

Treatment of the same original structures in H_2 resulted in formation of nanoparticles at temperatures lower than that observed when the atmosphere was He. No 6-atom particles were formed in H_2 but the range of Au-Au coordination in H_2 was the same as that observed in He.

11.5 XRD. – The three-dimensional structure of polycrystalline solids is often elucidated from a consideration of the diffraction pattern formed by the reflections of incident x-rays. A certain periodicity in the sample (~10 nm) is needed to produce a diffraction pattern of sharp peaks therefore x-ray diffraction methods are not used often to describe the structure of 'thin' films formed on the surface of an amorphous oxide. However, we were successful in developing meaningful x-ray diffraction data for the decoration of amorphous Cab-O-Sil with the polynuclear metal complex cation $\{[O\text{-}CH_2CH_2\text{-}NR_2\text{-}Cu\text{-}\mu OH\text{-}]_6\text{-}M\}^{3+}$ that formed a monolayer.[19] The data, Fig. 19, shows that for loadings of complex less than a monolayer, 200 μmol/g silica (D), the XRD data were similar to the undecorated silica (E). However, as the loading of metal complexes approached the value for a theoretical monolayer, 250 μmol/g silica (C), a subset of diffraction lines appeared and grew with incremental loading in excess of the monolayer (B). Some of these reflections could be identified with those observed for the polycrystalline, metal complex (A) and from a consideration of these data; we could interpret these results as the formation of a thin film of metal complexes decorating the surface of the silica.

The success in using XRD for this system can be attributed to certain aspects of the structure of the polynuclear metal complex that makes it possible to visualize with PXRD the formation of the thin film. The complex is shaped like a

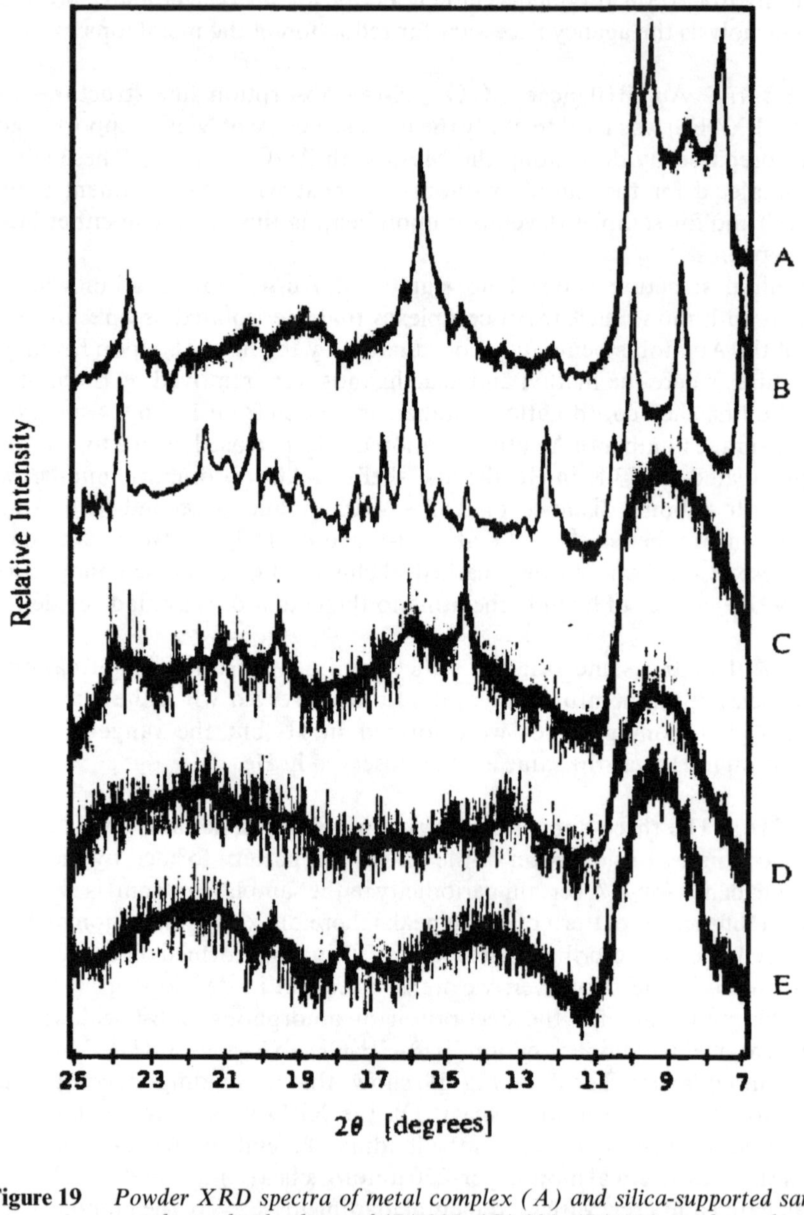

Figure 19 *Powder XRD spectra of metal complex (A) and silica-supported samples at metal complex loadings of (B) 285, (C) 221, (D) 166 mmol metal complex/g silica, and (E) silica. The intensities of spectra A & B are a factor of 10 greater than the intensities of spectra C-E*

disc, showing 3-fold inversion symmetry about the central metal ion, M. This simple structure permits the complexes to self-assemble into close-packed arrangement on the surface. Moreover, these cationic metal complexes bind to the surface as a result of ionic bonding and therefore do not need to form site specific bonds to atoms/species on the surface. This ionic bonding favors a self-assembly layer governed by the steric considerations of the metal complexes. Some of the reflections that appeared at monolayer loadings may not be present in the polycrystalline sample since the spacing on in the monolayer film may not be the same as the spacing realized in a 3-D structure.

XRD was not as useful for interrogating the arrangement of metal complexes in the surface film that formed from the site specific coordination bonds between surface siloxides and/or for complexes that showed lower symmetry [*e. g.*, *bis*-$M^{2+}(acac)_2$]. These metal complexes can arrange in the surface film in more than one configuration and thus give diffuse XRD diffraction patterns. The fortuitous combination of structural and bonding properties present in the hexameric copper metal complex are not present in a number of metal complexes and thus, the XRD technique has not been used as a general tool for interrogating the completion of a monolayer.

11.6 EPR. – Electron paramagnetic resonance has been used to describe the environment of some supported metal ions and with the most success when it is used to describe the fate of metal ions in metal complexes. Yamada reported one of the first studies using EPR to characterize *bis*-copper(II)acetylacetonate in X-type zeolite.[61] For samples having low loadings of Cu ($\sim$0.3 wt %), they reported the 4-line spectrum characteristic of isolated Cu(II) that demonstrates an axial field; however, for higher loadings of Cu metal complex in the zeolite, this distinctive EPR spectrum changed into one diffuse, featureless resonance. Thus, the information contained in the hyperfine structure of the resonance was forfeited at higher loadings of the metal, probably as a result of metal cation-metal cation interactions.

Choksi, also studied this complex on silica and observed the distinct, 4-line spectrum, characteristic of Cu^{2+} residing in an axial field at loadings of Cu(II) less than 1 wt%.[62] Silica-supported samples having higher loadings of this complex (2.0 wt%) did not have the well-defined hyperfine splitting as those samples at lower loadings and the 4-line spectrum was absent for those samples having loadings approaching a monolayer of the complex. He also reported EPR spectra for the following silica-supported samples: Cu-DETA (1.23 wt% and 3.0 wt% Cu/silica); Cu-bipy dimers $[(bipy)Cu\text{-}\mu OH\text{-}Cu(bipy)]^{2+}$ (0.44; 1.31, 2.27 wt% Cu).[62] While the Cu-DETA did not give well-resolved spectra, the Cu-bipy dimers did give the expected 4-line spectrum for loadings up to 2.27 wt% Cu(II) suggesting that there was no communication between the copper centers via the axial fields. These EPR spectra for supported copper samples can be interpreted using data for Cu(II)(benzac) diluted in $Pd(benzac)_2$ to show that isolated copper ions on the silica show spectra similar to isolated copper ions in 'solid solutions' with diamagnetic Pd(II).[35] One might speculate that interactions between the essentially flat, cation complex and the insulating silica surface prevented coupling of the copper ions through their axial fields since the silica is unlikely to support magnetic coupling interactions.

11.7 SQUID. – Superconducting quantum interference devices (SQUID) have been used to probe the magnetic properties of some metals and thereby permit some conclusions regarding the environment. Choksi, *et al.* employed SQUID to characterize the samples of $[(bipy)Cu\text{-}\mu(OH)_2\text{-}Cu(bipy)]^{2+}$/silica for their magnetic moments as a function of temperature and compared these results to the unsupported metal complex.[62] By careful analysis of the magnetic moment data for the supported and unsupported metal complexes, they showed that the supported metal complex demonstrated a small twist of the $Cu\text{-}\mu(OH)_2\text{-}Cu$ substructure when compared to the same part of the complex in the unsupported material. This small twist was responsible for the anti-ferromagnetic character demonstrated for the supported metal complex which was absent in the unsupported metal complex which showed ferromagnetic coupling of the Cu ions. They speculated that interactions between the surface siloxides and the dimeric metal complex were responsible for the deformation of the complex away from its intrinsically flat structure.

11.8 Selective Chemisorption. – The selective chemisorption technique can provide useful information regarding the morphology of the supported metal complex and for the decomposed metal oxide residue. We have used combinations of chemisorptions of selected probe molecules to characterize the surface Bronsted and Lewis acidity of acidic metal complexes. For example, we confirmed the expected bifunctional acidity in the supported metal complex $\{[O\text{-}CH_2CH_2\text{-}NR_2\text{-}Cu\text{-}\mu OH\text{-}]_6\text{-}M\}^{3+}$ by using ammonia as a probe for both Lewis and Bronsted acidity and by using nitric oxide as a selective probe of Lewis acidity[18, 19]. We also used these titrating agents to confirm the site densities and to measure the enthalpy of adsorption of both sites in the supported metal complexes when the central metal ion, M^{3+}, was changed from Al^{3+}, Cr^{3+}, and Fe^{3+}.[18, 19] This model bifunctional acid was especially useful in showing how the enthalpy of adsorption did not depend upon surface coverage for a particular metal complex (*i. e.*, M^{3+}) since all of the Bronsted sites demonstrated the same site energetics.[7]

When the surface metal ion is Cu, the combination of NO and N_2O is particularly useful for confirming that the monatomically dispersed Cu ions were separated by a distance of at least 4 Å.[11,63] Nitric oxide was used to count the total number of exposed Cu atoms and N_2O was used to count the number of Cu-Cu atom pairs that were separated by a distance of less than 4 Å.[62] We employed this battery of chemisorption tests to show that monatomically dispersed Cu/silica remained separated by at least 4 Å after repeated cycles of heating in oxidizing and reducing atmospheres.[7]

11.9 Chemical Reactions. – Probe reactions can be quite useful in the characterization of the surface condition of the supported metal oxide catalyst when there exist an unambiguous relationship between the property to be tested and the surface structure responsible for that property. Often, the issue of demanding versus facile reactions can be used to infer the ensemble size of the reactive site. We exploited this feature in the reaction of a small ester molecule, methyl acetate, to illustrate the effectiveness of our method for preparing supported Cu on silica catalysts.[64] The hydrogenolysis of these esters to form methanol and ethanol is generally thought to be a demanding reaction as well as the subsequent reaction of methanol to CO and H_2. However, the dehydrogenation of ethanol to acetaldehyde and the reverse reaction can be completed with high selectivity over isolated copper sites. We examined the demanding nature of this suite of probe reactions by preparing a series of supported Cu/silica catalysts, all showing a Cu loading ~ 2 wt%, but we varied the nuclearity of the metal complex precursor from 1, 2, 3, and 6 by using the following Cu metal complexes: $Cu(acac)_2$ as the mononuclear metal complex; $[(bipy)Cu\text{-}\mu(OH)_2\text{-}Cu(bipy)]^{2+}$ as the dinuclear metal complex; $\{[Cu(dapo)]_3\}^{3+}$ as the source of the trinuclear copper, and $\{[O\text{-}CH_2CH_2\text{-}NR_2\text{-}Cu\text{-}\mu OH\text{-}]_6\text{-}M\}^{3+}$ as the hexanuclear Cu complex. The hydrogenation of acetaldehyde to ethanol and the reverse reaction was catalyzed by all four catalysts, but only the catalysts derived from the mononuclear Cu complex showed 100% selectivity to the desired product. All other catalysts, having nuclearity > 1, showed other products such as ethyl acetate (a *trans*-esterification product). The methanol decomposition reaction did not occur over the catalysts prepared from mononuclear and dinuclear Cu sources, but it readily occurred over the two catalysts prepared from trinuclear and hexanuclear Cu sources. The hydrogenolysis of methyl acetate occurred only on the catalyst prepared from the hexanuclear Cu metal complex indicating that a large ensemble of Cu was required to catalyze this reaction. This study clearly demonstrated the power of the designed dispersion technique to prepare supported metal oxide catalysts with the desired catalytic properties merely by varying the type of precursor used in the synthesis.

12 Examples of Supported Metal Complexes Drawn from the Literature

We list in Table 2 a few of the systems that have been reported in the recent literature to show metal complexes supported on amorphous, non-porous oxides.[28,41,45,56,59–69] We also include a few recent reports on the decorating of mesoporous supports with metal complexes.[8,9,70] Complexes have also been introduced into porous, crystalline oxides[8,9,71] as well as placed on organic supports.[43,72] We reported the use of metal complexes as templates for forming familiar crystalline solids[73] and new crystalline materials, some of them adopting the chirality of the metal complex.[74,75] Preparations have appeared recently using dinuclear Pd(II) complexes $[Pd_2Me_2Cl_2(dppm)_2]$ as the precursor and these were reacted with a silica surface to produce the grafted dinuclear Pd complex with the elimination of methane from the complex.[76]

Table 2 *List of Recent Citations of Systems Using Metal Complexes*

Metal	*Amorphous Support*	*Precursor Complex*	*Reaction*	*Reference*
Cu/Ga/Zn	silica	$Cu(OCH_3)_2$ $Ga(acac)_3$ $Zn(acac)_2$	$H_2 + CO_2 \rightarrow CH_3OH$	65
Cu	silica	$Cu(en)_2$	$H_2 + H_3CCHO \rightarrow C_2H_5OH$	47
Ni	silica, alumina	$Ni(acac)_2$		66
Ni	α-alumina	$Ni(acac)_2$		17, 67
Pd	silica	$[Pd_2Me_2Cl_2(dppm)_2]$	6-aminohex-1-yne → 2-methyl-1,2-dehydropiperidine	76
Pd	silica, alumina	$Pd(acac)_2$	$NO + CO \rightarrow N_2 + CO_2$	40, 68
Mo/Cu/VO	silica, alumina	$MoO_2(acac)_2$ $Cu(acac)_2$ $VO(acac)_2$		45
Co	silica	$Co(acac)_3$		23
Fe	ZrO_2	$Fe(acac)_3$		57
Au	alumina	$Au(CH_3)_2(acac)$		60, 69
Ti, Co, Ni, Cr, Zn, Zr	silica	$Ti(acac)_4$, $Co(acac)_3$, $Ni(acac)_2$, $Cr(acac)_3$, $Zn(acac)_2$, $Zr(acac)_4$		16
Cr	alumina, silica	$Cr(acac)_3$		27
	Mesoporous			
Ru	MCM-41	$Ru(acac)_3$	H_2 + Prostaglandin Intermediates	71
VO	MCM-48	$O{=}V(O\textit{i}Pr)_3$	$H_3CH_2OH + O_2 \rightarrow HCOH + H_2O$	8, 9, 70
	Crystalline			
Cu	ZSM-5	$Cu(en)_2$	$NO \rightarrow N_2 + O_2$; MeOH → TriMe- + TetraMe-Benzenes	41
Cr	$AlPO_4$-5	$Cr(acac)_3$		28, 29
	Organic			
Cu, Co	activated C	$Cu(acac)_2$, $Co(acac)_2$	pinane + O_2 →2-pinane hydroperoxide	44
Cu, Mn, Fe, Ti	poly-benzimidazole resin	$Cu(acac)_2$, $Mn(acac)_3$, $Fe(acac)_3$, $TiO(acac)_2$	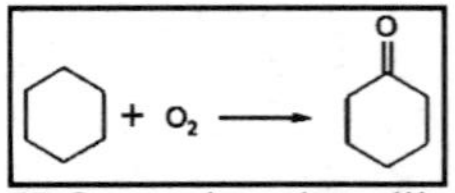	72
		Templates		
Co	zeolite A	$Co(en)_3$	H_2O_2 + phenol → dihydroxybenzenes	73
Co	$AlPO_x$	$Co(tn)_3$		74
Co	$AlPO_x$	$Co(dien)_3$		75

13 Summary

We showed here a brief and selected summary of the literature highlighting the recent publications on the subject of supported metal complexes. Where necessary, we have illuminated the science underpinning this technology and added fresh perspectives where appropriate. This technology appears to growing in popularity among academicians as a means to prepare supported metal oxides with control on the dispersion and nuclearity of the metal cations. Only recently, have papers appeared using more than one type of metal complexes so as to prepare mixed metal oxides. A few papers have appeared using polynuclear metal complexes as the precursor. Finally, no papers have appeared in print to show acceptance of this technique by the catalyst manufacturers.

14 Acknowledgments

The list of those to be acknowledged begins first and foremost with those who stood at the bench to do the work: the undergraduate students, the graduate students, the post-doctoral fellows, and the visiting academic and industrial scientists. Their devotion to duty and their intellectual gifts produced much of the work that originated from our laboratories. Their names appear in the listing of publications shown here and in the prior three reviews.

One person does deserve special recognition for his deep insight into the chemistry of polynuclear metal complexes and for his patience to share those insights with me: Professor Emeritus J. Aaron Bertrand, School of Chemistry and Biochemistry, Georgia Institute of Technology. The idea to use these metal complexes as model catalyst precursors originated from him. As a result of his prodigious abilities for synthesis of these systems, we had a ready inventory of fascinating metal complexes with which to decorate many surfaces. Much of the success in the early work was a result of our fruitful collaboration. We are indebted to him.

References

1. Ballard, D. G. H., *Adv. Catal.*, 1973, **23**, 173.
2. *Studies in Surface Science and Catalysis*, eds. Y. I. Yermakov, B.N. Kuznetsova, and V. A. Zakharov, Elsevier, Amsterdam, 1981, vol. 8
3. *Surface Organometallic Chemistry: Molecular Approaches to Surface Catalysis, NATO ASI series*, eds. J.-M. Basset, B. C. Gates, J.-P. Candy, A. Choplin, M. Leconte, F. Quignard, and C. Santini, Kluwer, Dordrecht, 1988, vol. **231**.
4. Guzman, J. and B. C. Gates, 'Supported molecular catalysts: metal complexes and clusters on oxides and zeolites', *Roy. Soc. Chem., Dalton Trans.* (2003), 3303-19.
5. Gates, B. C., *J. Mol. Catal. A*, 2000, **163**, 55.
6. Gates, B. C., in *Catalysis by Di- and Polynuclear Metal Cluster Complexes*, eds. R. D. Adams and F. A. Cotton, Wiley–VCH, Weinheim, 1998, p. 509.

7. White, Mark G., 'The Uses of Polynuclear Metal Complexes to Develop Designed Dispersions of Supported Metal Oxides', *Catal. Today*, vol. **18**, No. 1., pp. 73-109, (1993).
8. Van Der Voort, P., Mark B. Mitchell, Etienne F. Vansant, and Mark G. White, 'The uses of polynuclear metal complexes to develop designed dispersions of supported metal oxides part 1: synthesis and characterization', *J. Interface Science*, **5**, 179-208 (1997).
9. Van Der Voort, P., Michael Baltes, Etienne F. Vansant, and Mark G. White, 'The uses of polynuclear metal complexes to develop designed dispersions of supported metal oxides part 2: catalytic properties', *J. Interface Science*, **5**, 209-216 (1997).
10. Knozinger, H. and P. Ratnasamy, *Catal. Rev–Sci. Eng.* **17**(1), 31 (1978).
11. Kenvin, J. C., Ph. D. thesis, Georgia Institute of Technology, Atlanta, Ga (1988); Bertrand, J. A. and R. I. Kaplan, *Inorg. Chem.* **5**, 489 (1966).
12. Plyuto, Yuri V., Igor G. Babich, Lyudmila F. Sharanda, A. Marco de Wit, Johannes C. Mol, 'Thermolysis of $Ru(acac)_3$ supported on silica and alumina', *Thermochimica Acta* **335** (1999) 87-91.
13. Lofas, S., 'Methods for Site Controlled Coupling to Carboxymethyldextran Surfaces in Surface Plasmon Resonance Sensors', *Biosensors and Bioelectronics*, **10**, 813-22 (1995).
14. McKittrick, Michael W., and Christopher W. Jones, 'Toward Single-Site, Immobilized Molecular Catalysts: Site-Isolated Ti Ethylene Polymerization Catalysts Supported on Porous Silica', *JACS Communications* (2004) **126**, 3052 . . .
15. Venable, Margaret Hamm, 'Syntheses, structures and supported interactions of potential metal oxide catalyst precursors', MS thesis, Georgia Institute of Technology, (1990).
16. Leboda, R., J. Skubiszewska-Zieba and J. Rynkowski, 'Preparation and porous structure of carbon-silica adsorbents obtained on the basis of Ti, Co, Ni, Cr, Zn, and Zr acetylacetonates and acetylacetone', *Colloids and Surfaces A: Physicochemical and Engineering Aspects*, **vol 174**, Issue 3, 1 December 2000, pp 319-28.
17. Molina, R., M. A. Centeno, and G. Poncelet, 'α-Alumina-Supported Nickel Catalysts Prepared with Nickel Acetylacetonate. 1. Adsorption in the Liquid Phase', *J. Phys. Chem. B*. 1999, **103**, 6036-46.
18. Beckler, R. K., and M. G. White, 'Polynuclear Metal Complexes as Model Mixed Oxide Catalysts: Characterization', *J. Catal.*, **102**, pp. 252-256 (1986).
19. Beckler, R. K. and M. G. White, 'Polynuclear Metal Complexes as Model Mixed Oxide Catalysts: Selective Chemisorption of NH_3 and NO', *J. Catal.*, **109**, pp. 25-36 (1988); Beckler, R. K. and M. G. White, 'Polynuclear Metal Complexes as Model Mixed Oxide Catalysts: Isomerization Activity', *J. Catal*, **110**, pp. 364-374 (1988).
20. Coulier, L., V. G. Kishan, J. A. R. van Veen, and J W. Niemantsverdriet, 'Surface science models for CoMo Hydrodesulfurization Catalysts: the Influence of the support on hydrodesulfurization acidity', *J. Vac Sci. Technol. A*. **19**, Issue 4, 1 July/August 2001, pp 1510-5.
21. Coulier, L., V. H. J. de Beer, J. A. R. van Veen, and J W. Niemantsverdriet, 'Correlation between Hydrodesulfurization Activity and Order of Ni and Mo Sulfidation in Planar Silica-Supported NiMo Catalysts: the Influence of Chelating Agents', *J. Catal.* **197**, Issue 1, 1 January 2001, pp 26-33.
22. Haukka, S., A. Kytokivi, E. L. Lakomaa, U. Lehtoirta, M. Lindbladt, V. Lujala, T. Suntola, 'Preparation of Catalysts VI, Scientific Bases for the Preparation of Heterogeneous Catalysts', Poncelet, G., *et al*, Eds. Elsevier, Amsterdam, 1995, p. 957.

23. Blackman, L. B., A. Rautiainenb, M. Lindlad, O. Jylha and A. O. I. Krause, 'Characterization of Co/SiO_2 catalysts prepared from $Co(acac)_3$ by gas phase deposition', *Appl. Catal. A.: General*, vol **208**, pp. 223-34, (2001).
24. Lindblad, M., L. P. Lindors, and T. Suntola, *Catal. Lett.* 1994, **27**, 323.
25. Kytokivi, A., J. P. Jacobs, A. Hakuli, J. Merliainen, and H. H. Brongersma, *J. Catal.*, 1996, **162**, 190.
26. Kenvin, J. C., M. G. White and M. B. Mitchell, 'Preparation and Characterization of Mononuclear Metal Complexes as Model Catalysts', *Lang.* **7**, pp. 1198-1205, (1991).
27. Babich, Igor, V., Yuri V. Plyuto, Pascal Van Der Voort, and Etienne F. Vansant, 'Gas-phase deposition and thermal transformation of $Cr(acac)_3$ on the surface of alumina supports', *J. Chem. Soc., Faraday Trans.* (1997), **93**(17), 3191-6.
28. Plyuto, Yuri V., Igor V. Babich, Roger A. Sheldon, '$AlPO_4$-5 molecular sieve modified with $Cr(acac)_3$', *Appl. Surf. Sci.* **140** (1999), 176-81.
29. Sheldon, Roger A., Igor V. Babich, and Yuri V. Plyuto, 'The rational design of $Cr/AlPO_4$-5 molecular sieve', *Mend. Comm.*, Electronic Version, Issue 3, (1997), 94.
30. DiBella, S., G. Lanza, and, I. Fragala, '*Ab Initio* MO Study of the Molecular Structure, Vibrational Frequencies and Bond Dissociation Energy of *Bis*(2,4-pentanedionato-*O,O*')oxovanadium(IV)', *J. Chem. Soc.. Faraday Trans.* 1995, 91(17), 2709-14.
31. Burton, N. A., I. H. Hillier, M. F. Guest and J. Kendrick, *Chem. Phys. Lett*, (1989), **155**, 195; Hirota, F., and S. Shibata, *J Mol. Struct.* (*Theochem*), (1986), **137**, 373.
32. Hon, P-K., R. L. Belford and C. E. Pfluger, *J. Chem. Phys.* (1965), **43**, 3111.
33. Forsyth, G. A., D. A. Rice, and K. Hagen, *Polyhedron*, 1990, **9**, 1603.
34. Starikova, Z. A., and E. A. Shugam, *Struk. Khim.*, **10**, 290 (1969)
35. Hitchman, M. A., and R. L. Belford; 'The Electron Spin Resonance Spectrum of Copper-Doped Palladium *bis*-Benzoylacetonate Crystals', Electron Spin Resonance Metal Complexes, Proceedings Symposium (1969) pp 97-109.
36. Mitchell, Mark B., Vasumathi R. Chakravarthy, and Mark G. White, 'Interaction of Single and Multiple Layers of Copper Acetylacetonate [$Cu(II)(acac)_2$] Complexes with Silica Surfaces: An Infrared Diffuse Reflectance Study', *Langmuir*, 1994, **10** 4523-4529.
37. Pearson, R. G., H. B. Gray, F. Basolo, *J. Am. Chem. Soc.* (1960), **82**, 787; Person, R. G. and D. A. Johnson, *J. Am. Chem. Soc.* (1964), **86**, 3983.
38. Van Veen, J. A. R., P. C., de Jong-Versloot, G. M. M. van Kessle, F. J. Fels, *Thermochim. Acta* (1989), **152**, 359.
39. Van, Veen, J. A. R., G. Jonkers, W. H. Hesselink, *J. Chem. Soc. Farad. Trans.* (1989), **85**, 389.
40. Daniell, W., H. Landes, N. E. Fouadl and H. Knozinger, 'Influence of pretreatment atmospheres on the nature of silica-supported Pd generated via decomposition of $Pd(acac)_2$: an FTIR spectroscopic study of adsorbed CO', *J. Mol. Catal. A: Chemical*, **vol 178**, Issues 1-2, 23 January 2002, pp 211-8.
41. Rao, S. N. R., Ph. D thesis, Georgia Institute of Technology, Atlanta, GA (1997); Rao, S. N. R., and M. G. White, 'The Reaction of Methanol to Hydrocarbons: Effect of Adding Cu^{2+} to ZSM-5 Using $[Cu(en)_2]^{2+}$ by Ion Exchange', *Catalysis of Organic Reactions*, Ed. by Frank E. Herkes, **vol 75**, p. 63 (1998).
42. Rao, S. N. R., M. S. thesis, Georgia Institute of Technology, Atlanta, GA (1993).
43. Rao, S. N. R., E. Waddell, M. B. Mitchell, and Mark G. White, 'Selective Sulfur Dioxide Adsorbents Prepared from Designed Dispersion of Group IA and IIA Metal Oxides on Alumina', *J. Catalysis* **163,** 176-185 (1996).

44. Valente, A., A. M. Gotelho do Rego, J. J. Reis, I. F. Silva, A. M. Ramos, and J Vital, 'Oxidation of pinane using transition metal acetylacetonate complexes immobilized on modified activated carbon', *Appl. Catal. A*; *General*, **vol. 207**, Issues 1-2, 1 February 201, pp 221-8.
45. Baltes, M., O. Collart, P. Van Der Voort, and E. F. Vansant, 'Synthesis of Supported Transition Metal Oxide Catalysts by the Designed Deposition of Acetylacetonate Complexes', *Langmuir* (1999), **15**, 5841-5.
46. Bruce, D. A, J. A. Bertrand, M. G. White, 'The Effects of Molecular Structure upon Complex-Support Interactions', *Am. Inst. Chem. Eng. J.*, (1993), **39**, 1966.
47. Bruce, D. A., Ph. D. thesis, Georgia Institute of Technology, (1994).
48. Choksi, H. N., 'Silica Supported Binuclear Copper Catalysts: Preparation and Characterization', MS thesis, Georgia Institute of Technology, (1994).
49. Toupance, T., M. Kermarec, and C. Louis, 'Metal Particle Size in Silica-Supported Copper Catalysts. Influence of the Conditions of Preparation and Thermal Pretreatments', *J. Phys. Chem. B*, **104**(5), 965-72 (2000).
50. Babich, I. V., Y. V. Plyuto, P. Van Der Voort, and E. F. Vansant, 'Thermal Transformations of Chromium Acetylacetonate on Silica Surface', *J. Coll. Interfac. Sci.*, **189**, 144-50 (1997).
51. Plyuto, Y. V., I. V. Babich, L. Sharanda, A. Marco de Wit, J. C. Mol, 'Thermolysis of $Ru(acac)_3$ supported on silica and alumina', *Thermochemica Acta* **335** (1999) 87-91.
52. Mapes, J. E. and R. P. Eischens, 'The Infrared Spectra of Ammonia Chemisorbed to Cracking Catalysts', *J. Phys. Chem.* **58**, 1059 (1954).
53. Van Veen, J. A. R., G. Jonkers, W. H. Hesselink, *J. Chem. Soc. Faraday Trans.* **1** (85) (1989) 359.
54. Van Veen, J. A. R., P. C. de Jong-Versloot, G. M. M. van Kessel, F. J. Fels, *Thermochim. Acta.* **152** (1989) 359.
55. Van Veen, J. A. R., *Z. Phys. Chem. Neue Folge* **162** (1989), 215.
56. Kytokivi, A., A. Rautianin, A. Root, *J. Chem. Soc. Faraday Trans.* **93** (1997), 4079.
57. Van Der Voort, P., R. van Welzenis, M. de Ridder, H. H. Brongersma, M. Baltes, M. Mathieu, P. C. van de Veen, and E. F. Vansant, 'Controlled Deposition of Iron Oxide on the Surface of Zirconia by the Molecular Designed Dispersion of $Fe(acac)_3$: A Spectroscopic Study, *Langmuir*, (2002), 18, 4420-5.
58. Van Hengstum, A. J., J. G. van Ommen, J. G. Bosch and P. J. Gellings, *Appl. Catal.* **8**, (1983), 207.
59. Morey, M., A. Davidson, H. Eckert, and G. Stucky, 'Pseudotetrahedral $O_{3/2}V{=}O$ Centers Immobilized on the Walls of Mesoporous, Cubic MCM-48 Support: Preparation, Characterization and Reactivity toward Water as Investigated by ^{51}V NMR and UV-Vis Spectroscopy's', *Chem. Mater.* (1996), **8**(2), 486-92.
60. Guzman, J. and B. C. Gates, 'Structure and Reactivity of a Mononuclear Gold-Complex Catalyst Supported on Magnesium Oxide', *Angew. Chem. Int. Ed.*, **vol 42**, Issue 6, pp 690-3.
61. Yamada, Y., 'Electron Spin Resonance Studies of Bis(acetylatonato)copper(II) Adsorbed on X-type Zeolite. I. Interaction with Ammonia and Aniline', *Bull. Chem. Soc. Jpn*, **vol 45**, 60-3 (1972)
62. Choksi, H. N., A. Zippert, J. A. Bertrand, P. Berdahl, M. B. Mitchell, D. L. Perry, & M. G. White, 'A Copper/Silica Catalyst Prepared from a Monolayer Film of Copper 2,2'-Bipyridine Dimer on Cab-O-Sil', *J. Molecular Catalysis. A*, *Chemical*, vol 97, Issue 2, pp 85-99 (1995).
63. Kenvin, J. C. and M. G. White, 'Selective Chemisorption and Oxidation Reduction Kinetics of Supported CuO Prepared from Copper Acetylacetonates on Cab-O-

Sil', *J Catal.* **130,** pp. 447-458, (1991).
64. Bertrand, J. A., Dave Bruce, Homiyar Choksi, and Mark G. White, 'Supported CuO Catalysts Prepared from Polynuclear Metal Complexes', Symposium on Catalytic Selective Oxidation Presented before the Division of Petroleum Chemistry, Inc., American Chemical Society, Washington, DC, August 23-28, 1992.
65. Toyira, J., P. Ramirez de la Piscinal, J. L. G. Fierrob, and N. Homs, 'Catalytic performance for CO_2 conversion to methanol of gallium-promoted copper-based catalysts: influence of metallic precursors', *Appl. Catal. B: Environ.*, **vol 34**, Issue 4, 28 November 2001, pp 255-66.
66. Babich, I. V., Y. V. Plyuto, A. D. Van Langeveld, J. A. Moulijn, 'Role of the support nature in chemisorption of $Ni(acac)_2$ on the surface of silica and alumina', *Appl. Surf. Sci.* **115** (1997), 267-72.
67. Molina, R., and G. Poncelet, α-Alumina-Supported Nickel Catalysts Prepared from Nickel Acetylacetonate: A TPR Study, *J. Catal.* **173**, 257-67 (1998).
68. Pisanu, A. M., and C. E. Gigola, 'NO decomposition and NO reduction by CO over Pd/Al_2O_3', *Appl. Catl. B: Environ.*, **vol 20**, Issue 3, 8 March 1999, pp 179-89.
69. Guzman, J. and B. C. Gates, 'Reactions of $Au(CH_3)_2(acac)$ on γ-Al_2O_3: Characterization of the Surface Organic, Organometallic, Metal Oxide and Metallic Species', *Langmuir* (2003), **19**, 3897-03.
70. Coman, S., V. I. Parvulescu, B. Tesche, H. Bonnemann, J. F. Roux, S. Kaliaguine, and P. A. Jacobs, 'Stereo-controlled hydrogenation of prostaglandin intermediates over Ru-MCM-41 catalysts', *J. Mol. Catal. A: Chem*, **vol 146**, Issues 1-2, 20 October 1999, pp 247-56.
71. Plyuto, Y. V., I. V. Babich, R. A. Sheldon, '$AlPO_4$-5 molecular sieve modified with $Cr(acac)_3$', *Appl. Surf. Sci.* **140** (1999), 176-81.
72. Olason, G. and D. C. Sherrington, 'Oxidation of cyclohexane by *t*-butylhydroperoxide and dioxygen catalyzed by polybenzimidazole-supported Cu, Mn, Fe, Ru, and Ti complexes', *Reactive and Functional Polymers*, **vol. 42**, Issue 2, 15 November 1999, pages 163-72.
73. Bruce, D. A., Bertrand, J. A., Dai, E., Occelli, M. L., Petty, R., and Mark G. White, 'The Design of Microporous Crystalline Solids Using Polynuclear Metal Complexes', *Catalysis of Organic Reactions*, ed. by M. Scaros, Marcell Dekker, Inc., New York, pp. 545-52, (1994).
74. Bruce, D. A., Angus P. Wilkinson, Mark G. White, J. Aaron Bertrand, 'The Synthesis and Structure of a Chiral Layered Aluminophosphate containing the Template $Co(tn)_3^{3+}$', *J. Chem. Comm., Royal Society of Chemistry Journal*, 2059-2060, (1995).
75. Bruce, D. A., A. P. Wilkinson, M. G. White, & J. A. Bertrand, 'The synthesis and characterization of an Aluminophosphate with chiral layers, *trans*-$Co(dien)_2$ $Al_3P_4O_{16}$ $3H_2O$', *J. of Solid State Chemistry*, **125**, 228-233 (1996).
76. Richmond, M. K., S. L. Scott, G. P. A. Yap, and H. Alper, 'Preparation of Dinuclear Phosphine-Bridged Palladium (II) Species and Their Silica-Bound Analogues as Catalysts for the Cyclization of Aminoalkynes', *Organometallics*, **21** (16), 3395 (2002).

4
New Opportunity for HC-SCR Technology to Control NO_x Emission from Advanced Internal Combustion Engines

BY MOON HYEON KIM AND IN-SIK NAM

1 Introduction

Diesel and lean burn engine vehicles are much more attractive than conventional spark ignition ones, since they offer enhanced fuel economy and engine durability. However, to meet future emission standards a fundamental technology for reducing the emissions from both engines such as NO_x and particulate matter (PM) is currently required. There is no doubt that diesel and lean burn engines are the standard models for development of the recent advanced internal combustion engine. Although combustion measures and modifications of the engine have been widely employed so far, these approaches will no longer be appropriate for reducing the emissions from both engines to meet the future NO_x regulations.[1] It is widely agreed that an integrated approach for an advanced internal combustion system and advanced exhaust aftertreatment one should be simultaneously considered to meet such emission levels. This may be the reason why the installation of an aftertreatment system to the engines is unavoidable and mandatory in the near future. Commercially proven three-way catalytic converters (TWC) have been widely employed to remove NO_x in gasoline-fueled automotive exhausts; however, there are a few limitations in their commercial application. TWCs consisting of active noble metals, including Pt, Pd, and Rh, are hardly transposed for the reduction of NO_x from diesel and lean burn (LB) gasoline engines due to the huge amounts of excess O_2 contained in the exhaust stream, which can easily transfer the rhodium to an inactive phase, rhodium oxide, during the course of the reaction.[2] This implies that a new aftertreatment technology should be developed to meet the future NO_x regulatory requirements.

A great effort is underway to develop reliable aftertreatment systems for lowering NO_x emissions from diesel and LB engines. A variety of approaches have been proposed for NO_x aftertreatment of advanced vehicles including lean NO_x catalysts (LNC), NO_x storage and reduction (NSR) catalysts, selective catalytic reduction with urea (urea-SCR), and plasma-assisted catalysis (PAC).[1,3] Lean NO_x catalysts are mainly designed to reduce NO_x with unburned hydrocarbons already included in the exhaust stream in the presence of O_2 but result in

Catalysis, Volume 18

deNO_x efficiencies of lower than 50% and significant amounts of N_2O formation during the course of the reaction. The NSR catalysts were developed to remove NO_x from lean-burn gasoline engines operating at cyclic fuel-lean and -rich modes. Such NSR catalysts are still struggling for application to diesel engines because of the weak sulfur tolerance of the catalysts[4,5] by the sulfur currently contained in diesel fuel. The urea-SCR technology employs urea as a reductant for NO_x removal reaction, but the process produces NH_3 slip and additional requirement of infrastructure to supply and distribute the urea.[3,6] Finally, the plasma-catalyst system utilizes the plasma to oxidize NO into NO_2 which then reacts with suitable reductant over catalyst; however, this plasma-assisted catalytic technology still comprises challenging tasks to resolve the formation of toxic byproducts and the catalyst deactivation due to the deposition of the organic products during the course of the reaction as well as to prepare cost-effective and durable on-board plasma devices.[3,7,8] Consequently, all these technologies are not yet appropriate for commercial application to diesel and lean burn engine exhaust.

SCR of NO_x by using hydrocarbons (HCs) as a reductant, represented by HC-SCR, is widely recognized as a potential technology for NO_x emission control of advanced internal combustion engines. Some practical engineering advantages can be achieved in HC-SCR processes to remove NO_x from those engines. The first such advantage no longer requires extra reductant, since raw fuels appropriate for reducing NO_x can be readily supplied into the engine exhaust stream. Another one is on-board application without any installation of additional infrastructure, unlike urea-SCR processes for diesel NO_x emission control. Finally, this HC-SCR technology has the possibility to simultaneously remove both NO_x and unburned and/or partially burned HCs from engine emissions. Regardless of these engineering advantages, there are challenges for directly applying HC-SCR processes to NO_x emission controls, including poor hydrothermal stability and sulfur tolerance in diesel exhaust.[9-11] Consequently, most investigations to date have been focused on the use of light HCs for NO_x reduction, although many patents and research papers have dealt with deNO_x reactions under simulated operations of actual engines from which very complicated constituents, mainly containing lighter HCs, are emitted.[12,13]

Since the HC-SCR deNO_x technology was successfully established with Cu-zeolites in the presence of excess oxygen,[14,15] this has received great attention as a promising NO_x emission control for oxygen-rich mobile sources. There have been intensive investigations for the selective reduction of NO_x by HCs with excess oxygen. Based on the earlier works, the SO_2 effect on NO removal activity was moderate, whereas H_2O resulted in serious catalyst deactivation even with small amounts. Not only would high catalytic activity be essential for commercial application, but hydrothermal durability and sulfur tolerances of the catalyst must also be required from the view of catalyst life. In the early 1990s, an extensive review of this research area, mainly concerning the catalysts and their activity for reduction, was made.[16,17] Smits and Iwasawa[18] and Adelman *et al.*[19] reviewed the reaction mechanism for the reduction of NO by HCs. Tabata *et al.*[20] summarized more than 200 patents and their applications in the present deNO_x

technology. And recently, Parvulescu *et al.*[21] published an extensive review of catalytic NO_x removal, including HC-SCR, NH_3-SCR and direct decomposition technology, in which the catalysts and their catalytic properties were mainly discussed.

Both TWCs for purifying conventional gasoline engine exhausts and NH_3-SCR processes for stationary NO_x emission controls have been generally recognized to be sensitive to the presence of SO_2 in the feed gas stream but not H_2O vapor; however, the most urgent breakthrough in commercializing HC-SCR technology for aftertreatment systems of internal combustion engines (ICE) is to understand the diverse role of H_2O in catalyzing the selective reduction of NO_x by HCs, thereby designing a better SCR catalyst tolerable to it. Thus, the present review will focus on these issues, particularly relating to the effect of H_2O on the catalytic reduction of NO_x by HCs over a variety of catalysts to improve water tolerance and hydrothermal stability from knowledge-based understanding of the predominant factors in determining it. For this purpose, recent studies, efforts and attempts on these present topics in the literature will be mainly covered. The adverse effect of SO_2 on deNO_x catalysis is not of concern in this review, although strong sulfur tolerance of SCR catalysts is essential for practical applications to ICEs. It should be noted that the sulfur content in automotive fuels, particularly on-road diesel fuel, would be lowered to less than 15 ppm in 2006. In addition, the rapid development of the recent or future ICEs may also reduce the emission of NO_x and enable the present HC-SCR technology work.

2 NO_x Emission Regulation

Tier II emission standards of the EPA in the United States will be phased in between 2004 and 2009.[21a] All vehicles are categorized by gross vehicle weight rating (GVWR) and application areas of the vehicles. Chassis dynamometer tests using the FTP-75 cycle and supplemental federal test procedure (SFTP) are applied to light-duty vehicles (LDV, passenger cars) and light-duty trucks (LDT) with GVWR below about 3.86 ton, while heavy-duty vehicles having GVWR above 4 tons are tested by using the US-HD cycle and supplemental emission test (SET). The respective GVWR up to *ca.* 4.54 and 6.36 tons for medium-duty passenger vehicles (MDPV) and heavy-duty diesel vehicles (HDDV) can be tested with either light-duty standards or heavy-duty ones. Manufacturers of the vehicles being tested on the FTP-75 cycle may choose Bin numbers as long as the corporate average meets the applicable interim or final NO_x standard.

All light duty vehicles (LDV) and light light-duty trucks (LLDT) less than *ca.* 2.72 ton GVWT eventually have to meet the stringent NO_x emission standards (*ca.* 0.04 g/km, corresponding to the Bins number 5) by 2009, although a corporate average emission standard (CAES) of about 0.18 g NO_x/km will be applied to those not required to meet the Tier II standards for 2004 to 2006.[21a] The heavy light-duty trucks (HLDT) may be required to meet 0.12 g NO_x/km as the CAES for the phase-in years of 2004 to 2007. However, all the HLDTs being produced in 2009 may be regulated under the same CAES as the LDV and

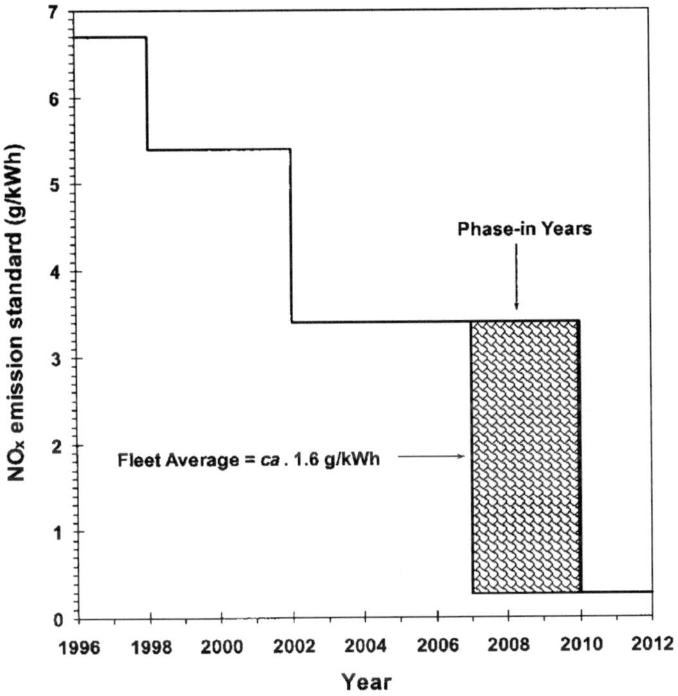

Figure 1 *NO_x regulation standards for heavy-duty diesel engines (adopted from ref. 22)*

LLDT, and the automotive manufacturers may use a mixed set of the Bin numbers of 1 (0.00 g NO_x/km) to 8 (0.12 g NO_x/km) to satisfy the CAES. Recently, the EPA also issued a major program to reduce emissions from heavy-duty vehicles, by 95%, compared to current NO_x standard of 5.36 g/kWh for both diesel and gasoline HDVs. The stringent Tier II emission standard of 0.27 g NO_x/kWh will be phased in between 2007 and 2010 for the HDVs (Figure 1).[22] Therefore, the diesel industry and related community are focusing on a way to meet the future NO_x emission standards.

3 Use of HCs for Catalytic Reduction of NO_x

A few '*classical*' studies on the reactivity of HCs to reduce NO_x with catalysts indicated that the use of such reductants for controlling mobile NO_x emissions was quite attractive to the automotive industry, thereby the advent of a new type of HC-SCR technology in the mid-1980s. An example may be the treatment process of the tail gas from nitric acid production plant *via* ammonia oxidation.[22a,23] The process includes the usual injection of excessive amounts of HCs over supported noble metals such as Pt, Pd and Rh to climinate the yellowish stack plume due to 0.1 – 0.5% NO_x, mainly NO_2, from the nitric acid plant.

Although the HCs are preferably consumed by 3 – 4% O_2 in the tail gas stream, they could play a significant role in converting the NO_2 to NO over the catalysts. Subsequently, the NO could be reduced on the catalyst surface through the following reaction.

$$(2x + \frac{y}{2})NO + C_xH_y \rightarrow (x + \frac{y}{4})N_2 + xCO_2 + \frac{y}{2}H_2O \tag{1}$$

From this stoichiometric reaction, it is suggested that NO from a variety of emission sources could be catalytically removed by using hydrocarbon species, if an appropriate catalyst were developed.

3.1 Challenges in HC-SCR Technology up to Early 1980s. – In this period, an attempt to use HCs as a possible reductant for the removal of NO_x in exhaust streams from stationary and mobile sources was established mainly with supported and unsupported metal oxides, in addition to supported noble metals.[24-28] Several catalysts could promote this type of NO_x reduction under appropriate conditions, among them being those consisting of these metal constituents.

Ault and Ayen[27] have utilized C_1-C_8 HCs including alkenes and alkanes for NO reduction over a Ba-promoted copper chromite to examine whether the chain length, electron deficiency, and structure of the HCs affect the catalytic NO reduction in the range of the reaction temperatures from 200 to 500°C. The general trend in that reaction revealed that an increase in carbon number in the HCs employed caused higher NO reduction efficiency at a given reaction temperature, and that for HCs containing the same carbon number, the electron-rich HCs showed the given performance of NO reduction at lower temperatures. Transition metal oxides such as MnO_2, Fe_2O_3, V_2O_5, Co_3O_4 and so forth have also been examined for NO reduction by using C_3H_6.[29] Among these catalysts, both Fe_2O_3 and MnO_2 were the most active catalysts for this reaction. Hu and Hightower[30] reported a comparative study of direct decomposition of NO and its reduction by CH_4 over a Pt/Al_2O_3 catalyst. Although both N_2 and N_2O were major products in the above simultaneous catalytic reactions, N_2O was predominantly formed for the catalytic decomposition reaction, whereas the reduction reaction initiated with the major N_2O production decreased as the production rate of N_2 increased. It was proposed that the NO reduction over the Pt catalyst involves the presence of a surface-adsorbed N_2O species as a reaction intermediate. As for catalytic NO reduction by HCs, few, if any, kinetic and mechanistic studies have been done. The catalytic reduction of NO with CH_4 over an alumina-supported Rh could be well illustrated by a Langmuir-Hinshelwood mechanism,[31] which was consistent with a kinetic study for NO reduction by C_2H_4 over a silica-supported copper catalyst.[25] These kinetic results suggest that NO removal reaction with HCs could be easily inhibited by guest molecules such as H_2O and SO_2, if they preferably interacted with the surface of SCR catalysts compared to two major reactants.

Since all the investigations discussed above were conducted mainly either in the absence of O_2 or with excess amounts of NO_x and HCs in feed gas streams, such HC-SCR reactions have not been so attractive from the viewpoint of commercial application particularly to ICEs under LB conditions. Major challenges in the HC-SCR processes during this period included accomplishing a viable combination of catalyst and reductant for selectively reducing NO_x sufficiently diluted in an effluent gas stream containing excess oxygen. Consequently, this motivated the development of a new version of HC-SCR technology, which may be phrased as 'Advanced HC-SCR Technology' distinct from the HC-SCRs studied up to the early 1980s, to remove NO_x from stationary and mobile sources, particularly from automotive vehicles.

3.2 Initial Study on the Development of Advanced HC-SCR Technology in Germany. – In the mid 1980s, the leading automobile manufacturers in Germany were particularly focused on a way to lower HCs and CO emissions from conventional gasoline-fueled vehicles to meet the stricter emission standards. Use of air/fuel ratios (AFR) greater than the stoichiometric value of 14.7 could decrease the mileage emissions of HCs and CO, as can be observed in Figure 2. Particularly, engine operation in the LB region leads to minimum emission levels of CO and NO_x, although the emission of HCs by that operation is significantly increased. With frequently employed three-way catalysts, compounds such as CO and HCs can be easily oxidized by the excess oxygen present in the exhaust gas stream; however, special deNO_x catalysts were required for successful and widespread usage of such LB engines. At that time, the diesel engine industry also needed deNO_x catalysts to meet future diesel emission standards.

Because of such strong demand for advanced HC-SCR technology applicable to automobile NO_x emission controls, Volkswagen AG and Bayer AG in Germany examined a variety of zeolitic catalysts on which NO_x from ICEs reacts

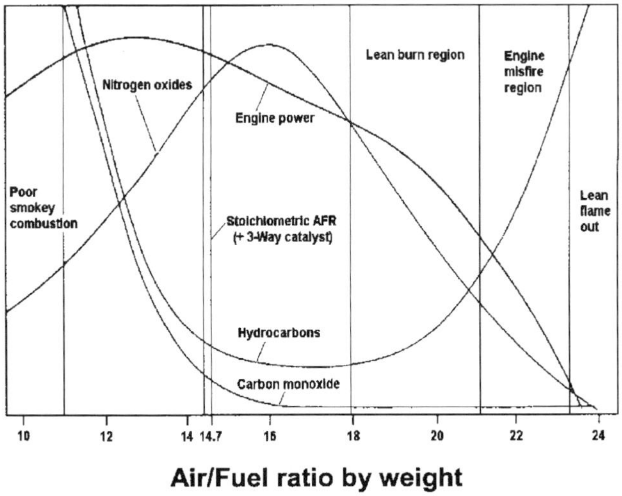

Figure 2 *Effect of air/fuel ratio by weight on emissions and engine power (adopted from ref. 16)*

effectively with HCs under the condition of containing excess oxygen. Transition metal ions exchanged in zeolites such as MOR, MFI and FAU (see Appendix A for the zeolite structure code) were quite active for NO reduction at 100 to 700°C of the reaction temperature and 10,000 to 50,000 h^{-1} of the reactor space time.[32-35] Cu-MOR catalyst exhibited 66% NO conversion within the temperature range from 150 to 230°C with C_2H_4 as a reducing agent in the absence of H_2O. Catalytic activity of 50% in terms of NO conversion at 350°C was observed for Cu-MFI catalyst, but other metal-exchanged MFI catalysts were relatively less active.[32] The capability of MFI zeolite to reduce NO at 300°C strongly depends on metal ions exchanged in the zeolite; for example, Ir-, Pt-, Rh-, Ni- and Co-MFI catalysts revealed lower NO_x conversion activity than Cu-MFI. FAU type zeolite exchanged with Cu, Fe, Mn, Ni, Co, Ag, V and Cr also revealed deNO_x performance with C_2H_4 and C_3H_6 but their NO removal activity was mainly less than 20% of NO conversion in the range of reaction temperature from 300 to 600°C. These findings were mainly obtained over a conventional monolith coated with the catalysts. It might be that Held and coworkers[32-35] at Volkswagen AG in Germany were the first to be aware of the present selective reduction technology using HCs for the removal of NO_x from automotive emission sources.

Held *et al.*[14] established the influence of H_2O on NO_x reduction activity of HC-SCR catalysts employed. Using Cu-MOR, 37 % of NO_x was removed at 350°C when the feed gas stream contained 1,000 ppm of NO_x in the presence of 40

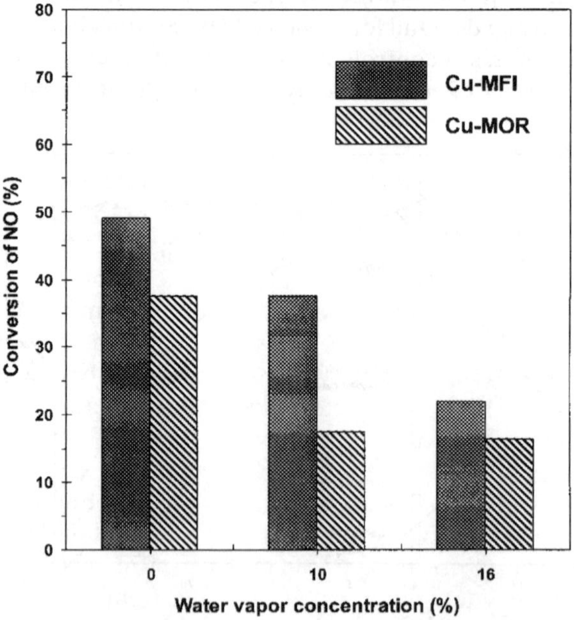

Figure 3 *Effect of water on NO_x conversion with zeolites. Reaction condition: NO_x 1,000 ppm, C_2H_4 40 ppm, O_2 1.5 %, GHSV = 13,000 and T = 350°C*[14]

ppm of C_2H_4 and 1.5% O_2; however, the NO_x conversion dropped to 17% when 10% H_2O was subsequently included in the feed gas stream, as shown in Figure 3. For Cu-MFI catalyst, the initial NO_x conversion of 50% fell to 37% in the presence of 10% H_2O and further decrease appeared with an increase in the feed concentration of water to 16%. The extent of the loss of NO removal activity for Cu-MFI catalyst in the presence of H_2O is somewhat milder than that over Cu-MOR. This indicates that the Cu-MFI is more tolerable than Cu-MOR in such a wet condition, being probably associated with the distinction of Si/Al ratio of both zeolite structures although this is not clear yet in the literature. It may have been the first open literature[14] for the adverse effect of H_2O on HC-SCR activity of Cu-exchanged zeolites, but the reason why those catalysts are severely deactivated by the presence of water vapor was not elucidated.

3.3 Pioneering Work for the Development of Advanced HC-SCR Technology in Japan. – In the research undertaken by Toyota Motor Corporation and Iwamoto and coworkers in Japan, an extensive screening test for the effective catalytic reduction of NO_x contained in vehicle engine exhausts was conducted at the end of the 1980s.[15,36-38] Fujitani *et al.*[36,37] of Toyota proposed a catalytic method for treating ICE exhaust gases by using Cu and Cu-based metals supported on Al_2O_3, SiO_2, SiO_2-Al_2O_3, zeolites and their physical mixture in the presence of HCs and excess oxygen. Selective removal of NO_x using HCs under a lean region of AFR was also investigated for a monolith reactor wash-coated with metal ion-exchanged zeolites by another research group at the same company.[38] Among the zeolite catalysts including transition metal ions such as Cu, Co, Cr, Ni, Fe, Mg and Mn, the copper-exchanged zeolites were the most preferable for the NO_x removal reaction. However, their catalytic activity depends on the structure of the zeolitic support and the concentration of HCs employed. Iwamoto and coworkers[15,39] found that Cu-MFI is the most active catalyst for NO_x reduction reaction; for example, its NO conversion activity is 80% at 500°C and 3000 h^{-1}, depending upon the Si/Al ratio and copper loading of the catalyst.

After these pioneering findings, the subsequent effect of O_2, CO_2, SO_2, H_2O, CO and HCs on the removal of NO was systemically examined to investigate if any alteration in the catalytic activity could be accompanied, since automotive engine exhaust truly contains a variety of gas components besides NO.[36-40] In the presence of O_2, the catalytic activity for NO removal reaction decreased, and a similar trend was also observed by the addition of either SO_2 or H_2O in the feed gas stream. However, the introduction of C_3H_6 to the gas stream caused a dramatic increase in the deNO_x efficiency, as shown in Figure 4, and similar catalytic behavior was observed when using CO instead of the HC. It is believed that HCs play a main role in the significant enhancement of NO removal activity even under wet oxidizing conditions, as acquired from the classical NO-HC reaction. The decomposition of NO over Cu-MFI catalyst in the presence of stoichiometric amounts of HCs as well as of excess O_2 could occur.[36-38] Thus, this NO removal reaction was later designated as not 'Decomposition' but 'Selective Catalytic Reduction'.[15] From this history for advanced HC-SCR process for NO_x

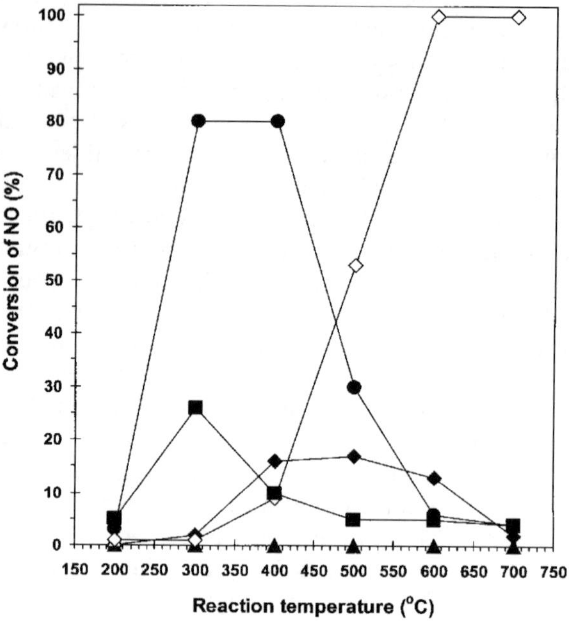

Figure 4 *Temperature dependence of catalytic activity for removal of NO over Cu-MFI-152 in various condition ($W/F = 0.3$ g.s/cm^3). (◆) NO (1,000 ppm); (▲) NO (1,000 ppm) + O_2 (1.0%); (◇) NO (1,000 ppm) + C_3H_6 (166 ppm); (■) NO (1,000 ppm) + O_2 (1.0%) + C_3H_6 (166 ppm); (●) NO (1,000 ppm) + O_2 (1.0%) + C_3H_6 (1,000 ppm)*[37]

abatement from ICEs, this technology came coincidently into the world 'during the survey of the effect of coexisting gases on the catalytic activity of copper ion-exchanged zeolites for the decomposition of NO'.[17]

A significant effect of H_2O vapor on NO_x reduction activity by C_3H_6 appeared for Cu-MFI-157 catalyst.[41] Introducing 3.9 % of H_2O to the feed gas stream containing 250 ppm of SO_2, the catalytic activity at 500°C suddenly decreased from 75 % of NO conversion to 40 %, as illustrated in Figure 5. The wet activity remained unchanged even for 1.5 h and was immediately restored to its initial dry conversion when H_2O was eliminated from the feed stream. This implied that the effect of H_2O was completely reversible and presumably suggested that the major active reaction sites on the catalyst surface, *i.e.*, Cu ions, may not be chemically altered by H_2O. However, the time on-stream hour is not so sufficient as to make it perceivable if any irreversible catalyst inactivation occurred at the given reaction temperature.

4 Catalyst and Reductant

4.1 HC-SCR DeNO$_x$ Catalysts. – Since Held and Konig[32], Held *et al.*[14,33-35] and Iwamoto[15] independently reported the selective catalytic reduction of NO_x by

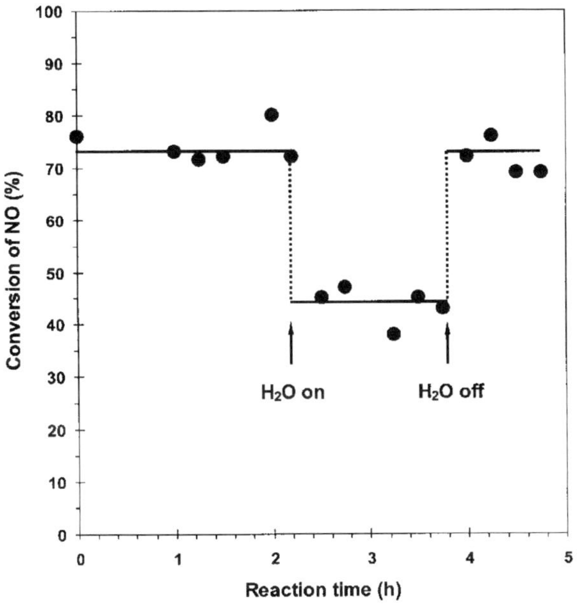

Figure 5 *Effect of H_2O on the catalytic activity of Cu-MFI-157 for the selective reduction of NO. [NO] = 600 ppm, [O_2] = 1.5%, [C_3H_6] = 940 ppm, [SO_2] = 250 ppm, [H_2O] = 3.9%, W/F = 0.1 g.s.cm^{-3} and T = 500°C*[41]

HCs over Cu-MFI catalyst on which a ppm level of NO_x was readily reduced by HCs even in the presence of excess oxygen, numerous catalysts listed in Table 1 have been proposed in the literatures to date. Representatively, transition metal ion-exchanged zeolites,[14,15,17,32,33-35,40-64] H-zeolites,[43,44,47-49,65-69] supported noble metals,[66,70-82] supported metals,[44,66,83-89] metal oxides,[44,70,90-92] solid acids[44,66,70,83,84,90] and perovskites[44,93] are well recognized to be active for HC-SCR reaction. Generally, transition metal-exchanged zeolites such as MFI, MOR, FER, LTL, BEA and FAU possess high deNO_x efficiency depending upon the kinds of metals and reductants employed. It is accepted that the selective reduction of NO_x by HCs in the presence of oxygen would be well catalyzed over transition metals including MFI and MOR type zeolite structures, especially MFI such as ZSM-5.[44,94] FER zeolite with Co ions is also one of the most promising support materials for the reduction of NO when reduced by CH_4.[95-99]

The efficiency of the above catalysts for NO reduction depends definitely on the kind of metals and their loadings onto supports, the type of reductants and the feed gas composition employed as well as on the kinds of supports and structure of the parent zeolite and its historical nature during preparation. In particular, the effect of the presence of H_2O and SO_2 in the exhaust gas from mobile sources is well documented on the maintenance of time-on-stream deNO_x activity of SCR catalysts, and their resistance to these co-existing gases is an essential parameter determining successful applications to engine sources. The durability of the documented catalysts under hydrothermal conditions should also be considered to verify if those were applicable to controlling vehicle NO_x

Table 1 *Literature-based representative catalysts for selective reduction of NO_x by hydrocarbons*

Zeolites and related materials
Bare zeolites[a]: MFI, MOR, FAU, FER, BEA, HEU, LTL, MFO
Metal ion-exchanged zeolites:
MFI with Cu, Fe, Co, Ce, Ga, Ag, Na, Zn, Ni, Mn, Mg, Mo, V, Cr, Ca, La, Pr, Nd, Ln, In, Ir, Pb, Pt, Rh, Ru and Pd
MOR with Cu, Fe, Co, Pt, Rh, Ru, Pd, Ti, W, Mo, Ce, Mg, Zr, Sn, Na, V, Cr, Ni, Zn, Ca, Ga, Sr, Ba and La
FER with Cu, Fe, Co, Cr, V, Zn, Pt, Pd, Mn and Ni
FAU with Cu, Co, Fe, Ce and Ga
LTL with Cu, Co and Fe
BEA with Cu and Co
HEU with Fe, Cr, Ni and Mn
Metallosilicates: Cu-, Fe-, Ga-, Al-, Co-, Ni-, Mn-, Mo-, Ti-silicate
Silicoaluminophosphates: Cu-, H-, Ca- and Pd-SAPO, H-MAPO, ALPO
Cordierites
Mullites
Metal oxides and related materials
Single metal oxides: Al_2O_3, SnO_2, TiO_2, ZrO_2, La_2O_3, Fe_3O_4 and Ag_2O
Mixed metal oxides: Al_2O_3-BaO, Al_2O_3-La_2O_3, ZnO-SiO_2, TiO_2-Al_2O_3, SiO_2-Al_2O_3, ZrO_2-TiO_2 and ZrO_2-Al_2O_3
Sulfate-promoted single and mixed metal oxides: Al_2O_3, TiO_2, ZrO_2, Fe_2O_3, ZrO_2-Al_2O_3 and ZrO_2-TiO_2
Metal oxides supported on the above metal oxides: Cu, Co, Ag, V, Ni, La, Mn, Ga, Cr, Ba, Ca, Sr, Mg, Zr, Cs, Sm, Mo, Ce and Fe
Perovskites
Transition metals
On zeolites: MFI, MOR, FER, MFO and FAU
On metal oxides: Al_2O_3, SiO_2, TiO_2, ZrO_2, La_2O_3, CeO_2, Cr_2O_3, ZnO, TiO_2-Al_2O_3, ZrO_2-TiO_2, TiO_2-ZrO_2-Al_2O_3, $AlPO_4$ and $AlBO_3$

[a] Zeolite structure codes designated by the Structure Commission of the International Zeolite Association (see Appendix A).

Table 2 *Literature-based hydrocarbons and related compounds for selective reduction of NO_x over SCR catalysts*

Hydrocarbons	
	Saturated HCs: CH_4, C_2H_6, C_3H_8, n-C_4H_{10}, i-C_4H_{10},
	Unsaturated HCs: C_2H_4, C_3H_6 and C_4H_8, C_5H_{12}, C_6H_{14}, C_7H_{16}, C_8H_{18}, C_9H_{20}, $C_{10}H_{22}$ and $C_{16}H_{34}$
Related compounds	
	Alcohols: CH_3OH, C_2H_5OH, C_3H_7OH and C_4H_9OH
	Common fuels: liquified petroleum gas, natural gas, diesel oil and gasoline
	Others: acetone, kerosene, dioxane, methylethylketone, toluene, benzene, xylene, ether, ether acetone, dimethylether, diethylether, formaldehyde, acetaldehyde, formic acid, acetic acid and methyl formate

emissions. Based on the water and sulfur tolerances of literature-known catalysts and their hydrothermal stability in these practical aspects, the most promising catalyst in HC-SCR technology for the deNO$_x$ reaction may be Fe-exchanged zeolite and supported Pt catalysts, depending on the type of the supports, although these state-of-the-art catalysts still entail several challenges such as engineering-unfriendly preparation technique, narrow operating temperature widow and undesirable product formation, as will be discussed later.

4.2 Hydrocarbons and Related Compounds. – A variety of HCs and related compounds have been employed for the reduction of NO$_x$, as summarized in Table 2. HCs used for NO$_x$ reduction in the presence of excess oxygen can be classified into two categories–selective and non-selective reductant–based upon the relative reactivity of HCs with NO and O_2 during the course of the reaction. Iwamoto and Hamada[40] have reported that C_2H_4, C_3H_6, C_3H_8 and C_4H_8 are selective for deNO$_x$ reaction while CH_4 and C_2H_6 are not selective, and a similar result has been observed for light hydrocarbons.[16] Both Cu-MFI and Al_2O_3 non-selectively catalyze NO$_x$ reduction by CH_4 and C_2H_6, but these HCs can be regarded as selective reductants if they are employed along with Co-MFI, Co-FER, Co-MOR,[46,51,95-97,99,100] Ga-MFI,[101,102] H-zeolite,[103] Pt/Al_2O_3,[104] Li-promoted MgO,[105] Pd-MOR[106] and In-FER[107] for the NO$_x$ removal reaction. Although the SCR activity is mainly related to the catalyst, the other variable may be the selectivity of the reductant for the present reaction system containing excess O_2 and water.[67,108] It means that the selectivity of HCs is primarily associated with surface features of the catalysts employed as extensively discussed.[109]

5 Deactivation of HC-SCR DeNO$_x$ Catalysts by Water Vapor

Most of the exhaust gas stream containing NO$_x$ from automotive engines includes H_2O in the concentration range of 2 to 18%; therefore, the strong water tolerance of deNO$_x$ catalysts is essential for its commercial application, in addition to their sulfur tolerance in the presence of SO$_x$ also contained in the exhaust stream besides NO$_x$. There have been efforts not only to elucidate the effect of H_2O on the deNO$_x$ efficiency of HC-SCR catalysts, but also to understand the reason why most of the catalysts significantly lose their activity during the deNO$_x$ catalysis in wet condition.[17,46,52-54,57,61,62,67,71-73,76,77,85-87,106,108,110-119] A few of the leading investigations have focused on how to improve the water tolerance of zeolite-based SCR catalysts for the reduction.[11,51,99] Based upon the earlier studies on the role of H_2O vapor in reducing NO$_x$ by HCs, this review intends to cover distinctive water tolerance of HC-SCR catalysts and their hydrothermal durability, as well as the approaches for improving the water tolerance of zeolite-based catalysts for the present NO reduction technology intensively developed over the last decades.

5.1 Water Tolerance of HC-SCR Catalysts. – Held and coworkers[14,32,33-35] first documented that both Cu-MFI and Cu-MOR could be significantly deactivated by H_2O during the reduction of NO_x using C_2H_4. Iwamoto *et al.*[41] obtained a similar result by examining the water tolerance of a Cu-MFI for HC-SCR reaction, as discussed already. This implies that the deactivation behavior of a variety of HC-SCR catalysts and the cause of the activity loss in the presence of H_2O are one of the most important challenges in this area of research. Thus, the response of the SCR catalysts to H_2O vapor contained in the feed gas stream is of particular interest in HC-SCR technology, as extensively summarized in Tables 3 to 8.

5.1.1 Fe-Exchanged Zeolites. Zeolite-based catalysts have not yet been applicable for NO_x from diesel and LB engines due to its poor time on-stream stability, especially in the presence of H_2O and SO_2, although their higher deNO_x activity than other SCR catalysts was extensively reported for the reduction with HCs. Recent progress in this technology, particularly regarding the method for suitable catalyst preparation, shows that Fe-exchanged MFI zeolites exhibit remarkable durability even under realistic engine exhaust conditions. Use of Fe-silicate as an SCR catalyst for NO removal reaction in the presence of C_3H_6 and excess O_2 has been reported by Kikuchi *et al.*[132] who observed its high deNO_x activity even with 10% O_2. Fe-zeolite catalysts, prepared by using a conventional ion exchange technique in an aqueous solution of the metal salts, have been also known to selectively reduce NO_x at a lean condition, significantly depending on the framework structure of the zeolites employed, as revealed in Figure 6.[44] Among the various types of the zeolite support, the Fe-MOR-97 catalyst exhibits the best SCR activity, which is greater than that for Fe-MFI-94 at all the reaction temperatures covered. The NO_x reduction efficiency of the Fe-MOR catalysts at 200°C is independent of the exchange level of Fe ranging from 50 to 150%. However, the water tolerance of either Fe based zeolite catalysts discussed has not been reported.

Feng and Hall[52,53] have proposed a unique method of preparing overexchanged Fe-MFI catalysts without the oxidation of Fe^{2+} to Fe^{3+} during the exchange and the formation of Brønsted sites. Not only could these catalysts be quite active for deNO_x reaction with i-C_4H_{10}, but they also exhibited unusual durability even in the presence of both 20% H_2O and 150 ppm SO_2 (Figures 7 and 8). These discoveries are contrasted on the well-known catalytic behavior of the most frequently-used Cu-MFI under concomitant H_2O and SO_2, particularly under high H_2O vapor pressure. The Fe-MFI-183 catalyst reveals 100% conversion of NO into N_2 at temperatures ranging from 450 to 550°C, while it is only possible to reduce about 55% NO over Cu-MFI-122 at the same temperature region, as compared in Figure 7. The Fe catalyst showed no activity loss even with 20% H_2O at 500°C, as shown in Figure 8 and Table 3, but the Cu-MFI-122 decreased from 53% to 5% in the wet stream. The achievable conversion of the Fe-MFI at the temperature does not decrease even for 500 h in the coexistence of the H_2O and SO_2. It is believed that the reaction temperature covered may be too high to evaluate the water tolerance. Regardless, it implies that the Fe-MFI is a

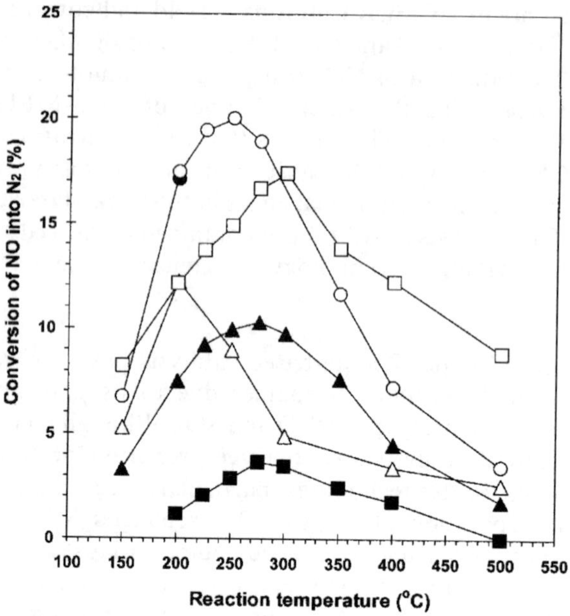

Figure 6 *Temperature dependence of the catalytic activities of iron ion-exchanged zeolites. (○) Fe-MOR-97; (●) Fe-MOR-58; (□) Fe-FER-49; (△) Fe-MFI-94; (▲) Fe-FAU-89; and (■) Fe-LTL-53. Catalyst weight, 0.5 g. NO (1,000 ppm); C_2H_4 (250 ppm); and O_2 (2%). Total flow rate, 150 cm^3/min[44]*

much more active and water-tolerable catalyst, compared to the Cu-MFI-122 as well as most of the other Cu-zeolites listed in Table 3. It should be noted that both catalysts consist of the same ratio of metal ion vs. aluminum on their surface. The NO_x reduction efficiency of Fe-MFI catalysts also depends on the Fe/Al ratios of the catalysts, as indicated from lower NO conversion for Fe-MFI-22 (Figure 7). This indicates that overexchanged Fe-MFI catalysts are preferable to NO_x removal reaction with HCs, at least i-C_4H_{10}, and tolerant to H_2O vapor at high temperature, which will be further discussed later.

Even with the same reaction conditions as those of Feng and Hall,[52,53] the extremely-high NO removal activity of overexchanged Fe-MFI catalyst claimed by these authors could not be reproduced by other research groups later.[54,55,140] Chen and Sachtler[55] tested dry and wet deNO_x activities of Fe-MFI catalysts, which had been prepared by using a wet ion exchange technique in the aqueous Fe solution, under identical conditions employed by Feng and Hall.[52,53] Their results may reveal that the de NO_x performance of such catalysts by HC-SCR reaction is quite sensitive to the subtle differences of the physicochemical characteristics of the parent zeolites employed. For instance, only a maximum NO_x conversion of 53 % was obtained for NO removal reaction with i-C_4H_{10} at 350°C over Fe-MFI-156 as shown in Figure 9, which is rather low deNO_x activity compared to the earlier results from the comparable Fe-MFI-183 catalyst prepared by Feng and Hall.[52,53] Even parent MFI zeolites containing virtually identical composition from different manufacturers reveal such distinctions in

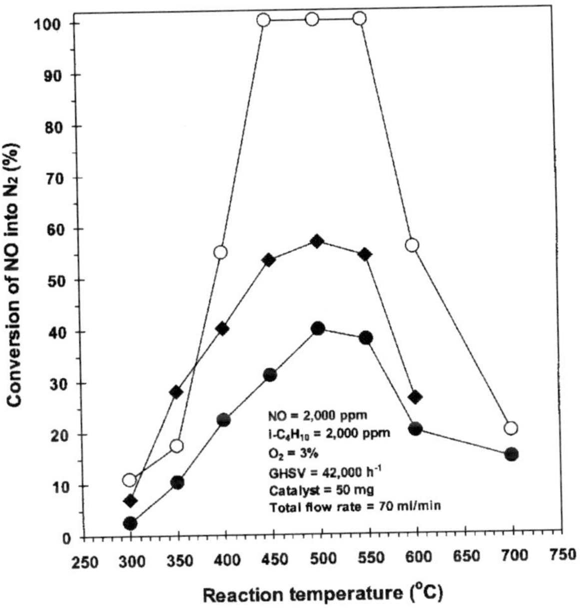

Figure 7 *Comparison of the conversion of NO into N_2 as a function of temperature for (●) Fe-MFI-22, (○) Fe-MFI-183 and (◆) Cu-MFI-122 (adopted from refs. 52 and 53)*

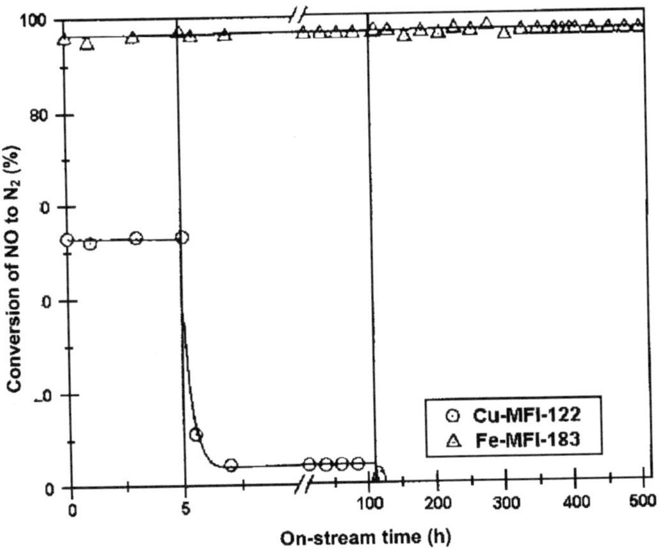

Figure 8 *Comparison of the effects of adding H_2O or H_2O and SO_2 to the gas stream over (△) Fe-MFI-183 and (⊙) Cu-MFI-122 at 500°C*[52]

Table 3 *Water tolerance of Fe-exchanged zeolite catalysts for the selective reduction of NO_x by HCs*

Catalyst[a]	*HC*	*H_2O (%)*	*Conversion of NO_x (%)*[b] *Reaction temperature (°C)* 250	300	350	400	450	500	550	*Ref.*
Fe-MFI-183	i-C_4H_{10}	20						97 (97)		52,53
Fe-MFI-300[c]	i-C_4H_{10}	10	13 (18)	34 (50)	76 (77)	51 (51)	36 (36)	27 (29)		55
	C_3H_8	10	20 (6)	60 (33)	56 (44)	44 (38)	33 (29)	26 (23)		57
Fe-MFI-225[d]	i-C_4H_{10}	0.7	15 (26)[e]	72 (68)[f]	62 (55)	37 (32)[g]				136
Fe-MFI[h]	i-C_4H_{10}	10	32 (21)	42 (37)	57 (52)	81 (49)	68 (37)	53 (28)	42 (24)	62

[a] Numbers next to the zeolite structure designate the exchange percentage of Fe ions.
[b] Numbers in parentheses indicate NO_x conversion at wet condition.
[c] Prepared by a chemical vapor deposition technique.
[d] Prepared by a solid-state ion exchange technique.
[e] At 275°C.
[f] At 325°C.
[g] At 425°C.
[h] Prepared by a mechnochemical ion exchange technique.

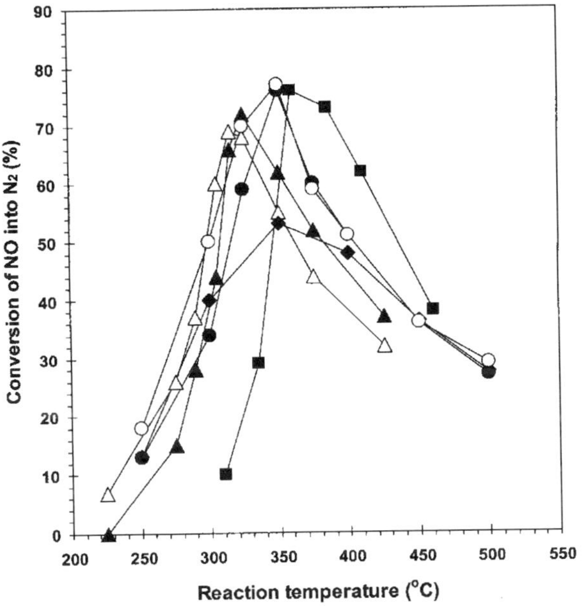

Figure 9 *NO_x reduction efficiency of overexchanged Fe-MFI catalysts prepared by using: (◆) a conventional ion exchange in an aqueous slurry;[55] (●,○,■) a sublimation technique;[55,139] and (▲,△) a solid-state ion exchange.[136] The feed gas composition for (◆,●,○,■) was: NO, 2,000 ppm; i-C_4H_{10}, 2,000 ppm; O_2, 3%; H_2O 0 or 10%; GHSV = 42,000 h^{-1}, and that composition for (▲,△) was: NO, 1,000 ppm; i-C_4H_{10}, 1,000 ppm; O_2, 3%; H_2O 0 or 0.7%; GHSV = 30,000 h^{-1}. (Closed symbols) in the dry feed stream; (open symbols) in the wet feed stream (adopted from refs. 55, 136 and 139)*

the NO_x reduction efficiency and its peak temperature, in addition to the peculiar reduction features of Fe ions in the zeolites as acquired by TPR profiles. Hall and coworkers[141] later supported these arguments in preparing overexchanged Fe-MFI catalysts, using a ferrous oxalate solution without air, on the fact that the catalyst is less inhibited by H_2O and hydrothermally more stable than other transition metal ion exchanged zeolites.

An alternative approach to overcoming these drawbacks for consistency of catalyst preparation was conducted by Chen and Sachtler,[54, 55] who proposed a way to obtain overexchanged Fe-MFI by subliming $FeCl_3$onto an H-MFI without contact with oxygen. The Fe-MFI-300 catalyst showed peak NO_x conversion of 76% for NO reduction with i-C_4H_{10} at 350°C, as also shown in Figure 9 and Table 3. The alteration of its catalytic activities at temperatures higher than 350°C was not observed, regardless of the presence of 10% H_2O in the feed gas stream as well as NO_x. Moreover, the presence of H_2O could promote deNO_x activity of this catalyst below 350°C reaction temperature, based upon their discussion; however, this may be associated with surface deposition of carbonaceous materials, at such low temperatures, which can be easily removed by water vapor, thereby apparently giving increased catalytic performance in wet conditions. This is quite evident from the fact that a gradual decrease in the catalytic reduction efficiency of the Fe-MFI at 300°C with respect to the time on-stream is perceived upon the use of C_4-HCs, including n-C_4H_{10} and i-C_4H_{10}, as a reductant for NO reduction in the absence of H_2O vapor, but not with C_3H_8 even at all of the reaction temperatures covered.[59] It is shown that the catalyst obtained *via* this preparation technique exhibits somewhat lower catalytic performance in the SCR reaction with *iso*-butane than that originally documented by Feng and Hall,[52] but the sublimation technique appears to be less sensitive to the historical nature of the parent zeolites, as indicated by the recent results from Battiston *et al.*[139] in Figure 9. In addition, these overexchanged Fe catalysts seem to be more effective when HCs containing carbon numbers greater than C_3 are employed as a reductant for NO removal reaction, as is evident from much lower catalytic activity in NO-C_3H_6-O_2 reaction as well as from no catalytic activity in CH_4-NO system.[57]

Another promising way to reproducibly prepare highly durable Fe-MFI catalysts to H_2O and SO_2 may be a solid-state ion exchange technique suggested by Cant and coworkers.[136] These authors did not use oxygen-free conditions during the catalyst preparation; however, the catalytic activity of Fe-MFI-225 for the selective NO reduction with i-C_4H_{10} in the presence of 3% O_2 was competitive to that obtained from the chemical vapor deposition method. Its NO_x conversion at 325°C appears to be 72% in the dry feed gas stream, as shown in Figure 9. In the presence of 0.7% H_2O, the dry activity almost maintains at the temperature, although both detrimental and beneficial roles for the catalytic deNO_x efficiency at all other temperatures are accompanied by H_2O. However, the wet durability of this catalyst under higher H_2O concentration than 10% has been unknown. Furthermore, catalyst stability as a function of time-on-stream in the presence of H_2O is questionable, since the ratio of Fe/Al = 1 for the catalyst was hardly achieved by using this solid-state ion exchange, regardless of the type and amount of Fe salts and the exchange conditions employed. Anyway, this technique does not require oxygen-free conditions during the catalyst preparation and can make it easy to prepare commercial Fe-MFI catalysts. In addition, significant amounts of CO were always produced over these Fe-zeolite catalysts during the course of the reaction, regardless of the preparation technique[54,55,57,136,140] which is undesirable for its practical applications.

5.1.2 Cu-Exchanged Zeolites. Among numerous NO_x reduction catalysts employed for the HC-SCR process, Cu-MFI is probably the most investigated catalyst for high temperature applications following the pioneering works from Germany and Japan, but unfortunately this catalyst still suffers from weak water tolerance and hydrothermal stability. Among a variety of such zeolite structures including MFI, MOR, LTL, FAU, BEA and FER, high NO removal activity by C_3H_8 has been observed over Cu-exchanged MFI and MOR zeolites, but not over FAU type zeolite.[121] In a wet stream with 14 % H_2O, the deNO_x conversion of Cu-MFI-87 decreased by 20 % at both 350 and 500°C and 40 % at 400°C, as listed in Table 4. The dry activity was not restored even after feeding the H_2O off,

Table 4 *Water tolerance of Cu-exchanged zeolite catalysts for the selective reduction of NO_x by HCs*

Catalyst[a]	*HC*	*H_2O (%)*	*Conversion of NO_x (%)*[b] *Reaction temperature (°C)* 350	400	450	500	*Ref.*
H-MOR	C_3H_6	7.3		84 (66)			48,49,67
	C_2H_4	7.3	66 (17)[c]				48,49,67
Cu-MFI	C_2H_4	10	50 (37)				14
	C_2H_4	16	50 (23)				14
Cu-MOR	C_2H_4	10	37 (17)				14
Cu-MFI-157	C_3H_6	3.9				75 (40)	41
Cu-MFI-122	i-C_4H_{10}	20				53 (4)	52,53
Cu-MOR-46	C_3H_8	5			24 (2)		120
Cu-MFI-87	C_3H_8	14	30 (8)	61 (19)	77 (45)	78 (55)	121
Cu-MOR-48	C_3H_6	7.3		78 (28)			48,49,67
	C_2H_4	7.3	65 (21)[c]				48,49,67
Cu-MOR[d]-44	C_3H_6	7.3		91 (85)			48,49,67
	C_2H_4	7.3	58 (34)[c]				48,49,67

[a] Numbers next to the zeolite structure designate the exchange percentage of Cu ions.
[b] Numbers in parentheses indicate NO_x conversion at wet condition.
[c] At 360°C.
[d] Natural zeolite consisting mainly of a MOR structure.

implying an irreversible activity loss by water vapor. This differs from the earlier results[41,48,49,67] but is in good agreement with the irreversible catalyst deactivation of a Cu-MFI-104 catalyst to remove NO from an actual diesel engine in the presence of 7% H_2O at 400°C.[142]

An extensive study of the loss of NO removal activity in the presence of H_2O has been conducted for Cu-MOR catalysts by C_2H_4 and C_3H_6.[48,49,67,112] Figure 10 shows the water tolerance of synthetic MORs such as H-MOR and Cu-MOR-48 and natural zeolite-based Cu-NMOR-44 for NO reduction by C_2H_4 as a function of the amount of H_2O contained in the feed gas stream. The activity of the synthetic MOR without Cu ions decreased from 66 to 17% under the wet stream containing 7.3% of H_2O; however, the water tolerance could have been appreciably improved as Cu ions were exchanged in the zeolite. The Cu-NMOR-44 revealed better water tolerance than the synthetic MORs containing Cu ions. If C_3H_6 was employed as a reductant, the catalytic activity loss of the Cu-NMOR-44 was less than 5% even at 16% of H_2O in the feed gas stream, as illustrated in Figure 11. The extent of the deactivation of the synthetic MOR catalysts is improved when C_3H_6 is used instead of C_2H_4, regardless of the presence of Cu ions; thus, C_3H_6 is a better reductant than C_2H_4 under a wet stream. The Cu-exchanged MOR type zeolite shows milder catalyst deactivation by water than the copper-free catalyst, and the natural zeolite-based Cu-NMOR particularly possesses stronger water tolerance than the synthetic MOR-based zeolites. It indicates that the zeolite structures and the existence of the copper ions are mainly associated with the distinctive water tolerance of the catalyst.

The NO removal activity of Cu-MOR-48 in the presence and absence of H_2O was completely reversible.[67] The reversibility of NO conversion has also been

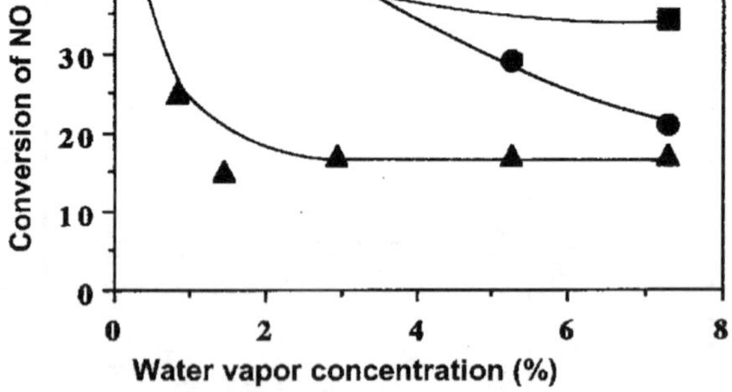

Figure 10 *Water tolerance of mordenite type zeolite catalysts for the reduction of NO by C_2H_4: (●) Cu-MOR-48; (▲) H-MOR; (■) Cu-NMOR-44. Reaction condition: NO 500 ppm, C_2H_4 1,000 ppm, O_2 4.2 % and $T = 360°C$*[112]

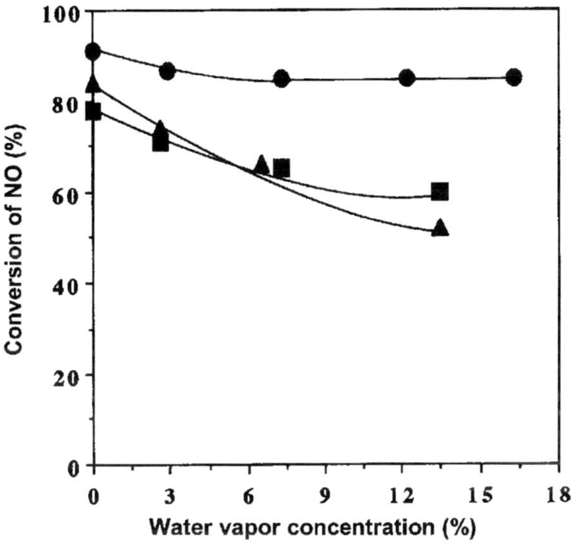

Figure 11 *Water tolerance of mordenite type zeolite catalysts for the reduction of NO by C_3H_6: (■) Cu-MOR-48; (▲) H-MOR; (●) Cu-NMOR-44. Reaction condition: NO 500 ppm, C_3H_6 2,000 ppm, O_2 4.2 % and T=400°C*[112]

observed for H-MOR and Cu-NMOR-44 catalysts,[49,67,112] regardless of reductant, which is consistent with the earlier studies for Cu-MFI-157[41] as well as for Co-FER-78 and Co-MFI-140.[46,97] Similar reversible behavior in the conversion of C_2H_4 and C_3H_6 into CO_x occurred for all the Cu-MOR catalysts during the cyclic injection of water into the feed gas stream. It can be anticipated that Cu ions associated with the exchange sites of zeolites are responsible for NO_x reduction activity; therefore, an irreversible deactivation would occur if the state of the Cu ions or the framework structure of the zeolite were changed, as published in earlier studies.[111,143] It suggests that the activity deterioration of the Cu-MOR type catalysts is probably associated with the adsorption characteristics of HCs and H_2O rather than due to the chemical alteration of the catalyst in the presence of water vapor.

5.1.3 Co-Exchanged Zeolites. Armor and coworkers[46,96,97] have examined the water tolerance of metal ion-exchanged MFI, FER and MOR zeolites for selectively reducing NO by CH_4. Both Co-MFI-98 and Co-FER-78 catalysts were highly active for the NO removal reaction, compared to Co-MOR-94, Mn-MFI-106, Ni-MFI-140 and H-MFI. However, the catalytic activity was still suppressed in the presence of 2% H_2O, regardless of the presence of metal ion exchanged in the catalyst (Table 5). As an example, Co-MFI-140 catalyst offered 47% of steady-state NO conversion without H_2O but the presence of 2% H_2O in a gas mixture resulted in an activity loss of more than 50 % at reaction temperatures such as 400 and 450°C, as shown in Figure 12. Such an adverse

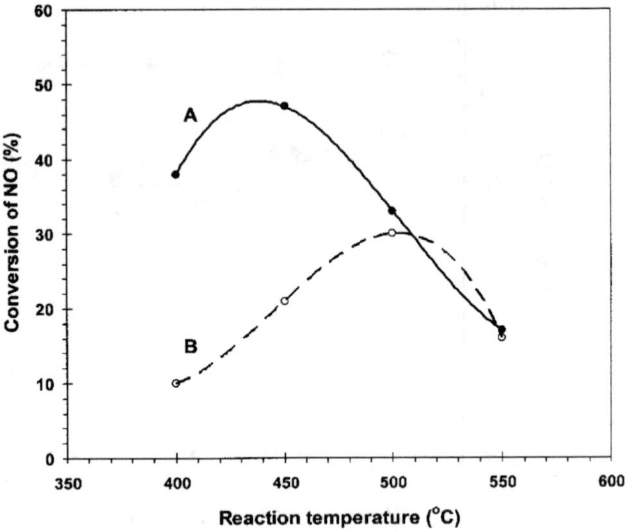

Figure 12 *Effect of water on the NO conversion over Co-MFI-140 as a function of reaction temperature. The reaction was run at GHSV=30,000, [NO]=820 ppm, [CH_4]=1,015 ppm, and [O_2]=2.5% ([H_2O]=2% for the wet feed). The reaction was run first with the dry feed with increasing temperature (line A), then 2% water vapor was added at 560°C and the reaction was run with decreasing temperature (line B). Data were collected at each temperature for a period 1 to 2 h, and only stable, average data are shown*[46]

H_2O effect on $deNO_x$ activity was completely reversible;[46,97] therefore, the alteration of cobalt state and zeolite structure may not be suspicious in the present system, either. Based upon the comparison of Co-MFI-98 catalyst with Co-FER-78 for NO_x reduction by CH_4, Li and Armor[97] observed that FER type zeolite exhibits the best NO removal activity even in a wet stream containing 2 % H_2O, as indicated in Table 5; however, this only occurs when CH_4 as a reducing agent is employed, since the $deNO_x$ activity of Co-BEA-80 catalyst under a wet stream with C_3H_8 was much greater than that of Co-FER catalyst.[144] The overall conversion of CH_4 significantly decreased under the wet condition as well, irrespective of zeolite structure employed. A complete reversible catalyst deactivation due to water was again observed for the both Co-exchanged zeolites during a cyclic inclusion of water into the feed gas stream. It represents that Co-zeolites may contain more opportunity for practical application than Cu-zeolites if the contents of the metal on the surface of both catalysts are comparable.

Among Co-FER, -MOR, -BEA and -MFI zeolites for NO reduction using CH_4, the most desirable catalyst for the CH_4-SCR reaction, if only their peak $deNO_x$ conversion is considered, has been shown to be a Co-FER within the reaction temperature range from 400 to 450°C, where an NO_x reduction efficiency greater than 80% has been observed. Such a narrow temperature window in the present SCR reaction is quite common for Co-zeolite catalysts,[97] while

Table 5 *Water tolerance of Co-exchanged zeolite catalysts for the selective reduction of NO_x by HCs*

			Conversion of NOx (%)[b]								
			Reaction temperature (°C)								
Catalyst[a]	*HC*	*H_2O (%)*	*250*	*300*	*350*	*400*	*450*	*500*	*550*	*600*	*Ref.*
H-MFI	CH_4	2					16 (5)	24 (2)			46
Co-MFI-140	CH_4	2					47 (23)	33 (31)			46
Co-MOR-94	CH_4	2					33 (25)	28 (25)			46
Co-MFI-98	CH_4	2						40 (29)	27 (28)	21 (22)	97
Co-FER-78	CH_4	2						60 (28)	50 (40)	40 (32)	97
Co-FER-80	CH_4	10		1 (1)	29 (1)	86 (12)	100 (35)	93 (66)	75 (62)	54 (44)	99
Co-Pd-FER-76[c]	CH_4	10		7 (0)	24 (11)	70 (24)	80 (45)	65 (59)	51 (51)	37 (35)	99
Co-MFI	C_2H_6	2					50 (22)				122
Co-FER-94	C_3H_8	9				ND (15)		ND (39)			144
Co-BEA-80	C_3H_8	9		ND (0)		ND (84)		ND (86)		ND (55)	144
Co-MFI-226[d]	i-C_4H_{10}	10	14 (9)	18 (13)	37 (65)	99 (92)	94 (86)	81 (84)			154
Co-MFI-196[e]	i-C_4H_{10}	10	9 (8)	12 (15)	46 (55)	56 (64)	57 (66)	44 (63)			154

Note. ND = no data.

[a] Numbers next to the zeolite structure designate the exchange percentage of Co ions.

[b] Numbers in parentheses indicate NO_x conversion at wet condition.

[c] Sum of each exchange percentage value for both Pd and Co (Pd/Al = 0.18 and Co/Al = 0.20).

[d] Prepared by a chemical vapor deposition technique.

[e] Prepared by a solid-state ion exchange technique.

In-FER catalyst revealed 50 – 60% of $deNO_x$ activity at much wider temperature ranges from 350 to 600°C.[51] Recently, the dependence of Co/Al ratio for Co-FER catalysts on NO removal activity by CH_4 has been reported by Nam and coworkers,[99] as shown in Figure 13. Co-FER-80 catalyst achieved $deNO_x$ conversion of 100% at 450°C under a dry condition but 55% for Co-FER-44 at the same temperature. However, Co-FER catalysts containing Co/Al ratios of 0.69 and 0.91 revealed lower dry activity than the Co-FER-80 at the reaction temperature. All the Co-FER catalysts possessed the same Si/Al ratio of 9. This agrees well with the dependence of catalytic CH_4-SCR activity on Co content over a Co-MFI catalyst.[98,145,146] In a wet stream containing 10% H_2O, the extent of the activity loss of the Co-FER-80 catalyst is shown to be about 65% at 450°C, but the catalytic activity of Co-Pd-FER-76, consisting of (Pd + Co)/Al ratio comparable to the Co-FER-80, decreases by 35% (Figure 13 and Table 5). If an increase in Co/Al ratio is possible for a catalyst containing a low ratio of Pd/Al, the extent of the suppression of the NO conversion by the H_2O becomes significantly improved, and it may be much better than Co-only ferrierite, as revealed for Co-Pd-FER-156 (Co/Al = 0.75 and Pd/Al = 0.032). It was proposed that Pd ions present in the divalent form probably participate in the oxidation of NO to NO_2, in addition to an increase in NO adsorption. This is, however, in contrast with earlier results: it has been reported that both well-dispersed Pd^{2+} and Co^{2+} ions are involved mainly in methane activation while oxo-Co species boost catalytic $deNO_x$ activity, depending strongly on parent

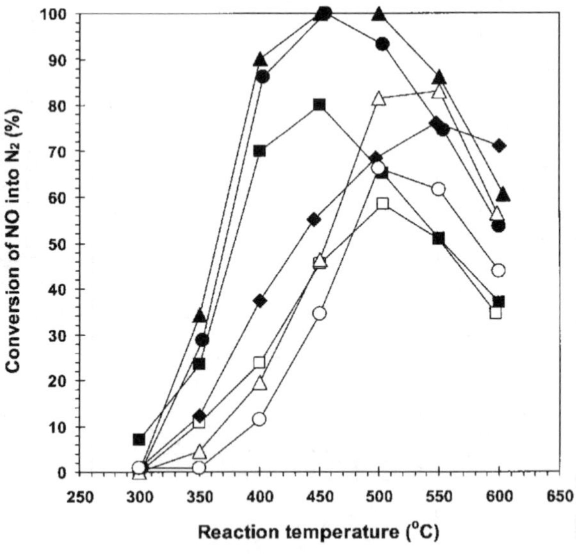

Figure 13 *Water tolerance of Co-only and Co-Pd-ferrierite catalysts for NO reduction with CH_4: (◆) Co-FER-44; (●,○) Co-FER-80; (■,□) Co-Pd-FER-75; (▲,△) Co-Pd-FER-156. Reaction condition: NO 1,200 ppm, CH_4 2,400 ppm O_2 2.6%, H_2O 0 or 10% and GHSV = 14,000 h^{-1}. Closed and open symbols represent the respective dry and wet $deNO_x$ conversions (adopted from ref. 99)*

zeolite structure, due to the oxidation of NO to NO_2.[110,113,147,150] Regardless, the cocation plays a positive role for improving the water tolerance of Co-FER catalysts with respect to the extent of the decrease of the peak NO conversion. Similar cocation effect on the catalyst water tolerance was observed in the earlier studies for Co-based second metal-exchanged zeolites such as Co-Pd and Co-Pt.[113,147-149] These results may not be sufficient to compete with the activity of overexchanged Fe-zeolites for practical applications to NO_x emission controls of LB gasoline and diesel engines. It is proposed that such Co-based zeolite catalysts may have a potential for commercial application to natural gas-fueled engines. It should be noted that CH_4 is no longer an active reductant over the Fe catalysts, as stated previously. However, hydrothermal durability has been unknown for these Co-based catalysts.

Metal-exchanged zeolites, particularly Co, have been widely considered to be far more active than ones in the form of H and Na without metal for CH_4-SCR reaction, although a synergetic role of Co and Brønsted acid sites was proposed by Regalbuto *et al.*[151] H-zeolites were active for the reaction but severely deactivated in the presence of H_2O due to its competitive adsorption on the acidic sites with the reactants.[46,113] The use of high Co-exchanged zeolite usually resulted in a remarkable increase for both the activity and the durability of Co-zeolites.[51,96,98,99,146,152] Moreover, Campa *et al.*[153] recently reported a negligible role of H sites remaining on the surface of Co-MOR catalyst for NO reduction with CH_4 in the presence of excess oxygen. It suggests that it is necessary to completely substitute all acidic proton sites present in zeolites into Co ions for obtaining a better catalytic performance in CH_4-SCR reaction. Conventional wet ion exchange technique has difficulty in preparing overexchanged Co zeolite catalysts. Other ways to increase the content of Co on the surface of zeolite catalysts include chemical vapor deposition,[154] solid-state ion exchange[155,156] and *in situ* hydrothermal synthesis technique,[157] in addition to a combination of wet ion exchange and impregnation.[158] Although solid-state ion exchange can prepare a catalyst containing Co/Al ratios greater than unity with no Brønsted acid sites on its surface, it may form non-isolated Co^{2+} species including $CoCl^+$ and $CoOH^+$which are inactive for CH_4-SCR reaction.[155] The CVD technique, consisting of subliming $CoCl_2$ onto an H-MFI under oxygen-free conditions, may prepare a Co-MFI with high Co/Al ratio, which is very active for NO_x reduction by i-C_4H_{10} even in the presence of 10% H_2O, as shown in Figure 14. The Co-MFI-226 reveals an activity loss of less than 10% at 400°C by water vapor and exhibits much better catalytic performance than Co-MFI-196 obtained by wet ion exchange (Table 5 and Figure 14). This indicates that the use of suitable catalyst preparation method can improve the water tolerance of Co-zeolites, besides the variation of the metals including bimetal catalyst such as Co-Pd and Co-Pt as discussed.

5.1.4 Other Metal-Exchanged Zeolites. Nitrogen oxides have been selectively reduced by HCs over zeolite catalyst exchanged with other metal ions including Pd, Pt, Rh, In, Mn, Ga and Ce, besides Fe, Cu and Co. Pd-zeolites[50,99,106,118,119,159-163] were most frequently investigated in this catalyst group,

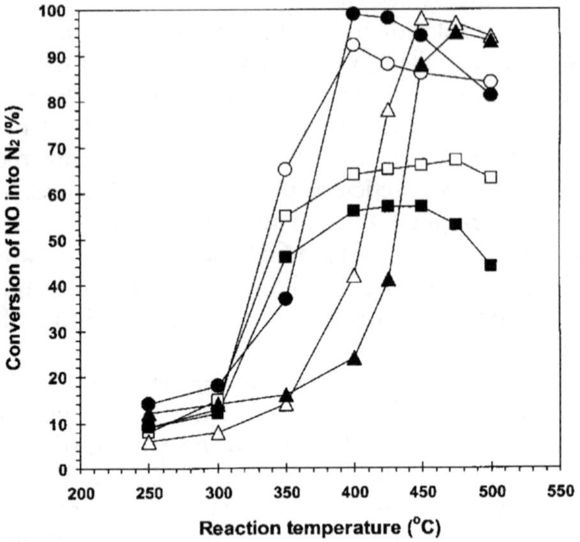

Figure 14 *NO removal activity of Co-MFI catalysts prepared by different methods: (●,○) chemical vapor deposition, Co-MFI-226; (■,□) solid-state ion exchange, Co-MFI-196; and (▲,△) wet ion exchange, Co-MFI-88. Reaction condition: NO 2,000 ppm, i-C_4H_{10} 2,000 ppm O_2 3%, H_2O 0 or 10% and GHSV = 42,000 h^{-1}. Closed and open symbols represent the respective dry and wet deNO_x conversions (adopted from Table 3 in ref. 154)*

and in the presence of H_2O their catalytic activity for NO_x removal reaction with HCs was significantly inhibited (Table 6). The extent of the activity loss of these catalysts under wet condition depends strongly on the structure of zeolite and its Si/Al ratio, and the temperature exposed to H_2O vapor.[106,118,119,161,162] Extensive characterization of the zeolite structure in relation to Pd-cation site elucidated the differences in activity among the various zeolites.[150,164,165] Ga-zeolites were highly active for the reduction of NO by CH_4 andC_3H_8.[101,166] In a comparative study with Co-MFI catalyst for the reduction of NO by CH_4, Ga-MFI-162 revealed higher performance of NO removal activity and was more selective than Co-MFI catalyst, but it showed more severe activity loss in the presence of H_2O. The respective deNO_x conversions for Ga-MFI-162 and Co-MFI at 500°C fell from 40 to 13 % and from 40 to 35 % in the presence of 2% H_2O.[166]

Pt-zeolite catalysts were also employed for the selective reduction of NO_x by HCs[73,167] and were very active at lower reaction temperatures compared to Fe-, Cu- and Co-zeolites, but little, if any, literature was found for the Pt-exchanged zeolite catalysts dealing with the water tolerance. Iwamoto and coworkers[168,169] conducted a comparative investigation of Pt- and Cu-MFI and Fe-MOR for their performance in NO reduction by C_2H_4. Pt-MFI-97 was found to be more active than the other two metal-exchanged zeolite catalysts at low temperatures and its activity for NO conversion at 200°C is hardly affected by the addition of 8.6% H_2O into the feed gas stream (Table 6), whereas the conversion of N_2O formed during the course of the reaction slightly decreases by less than 10%. In

Table 6 *Water tolerance of other metal-exchanged zeolite catalysts for the selective reduction of NO_x by HCs*

			Conversion of NO_x (%) [b]						*Ref.*
			Reaction temperature (°C)						
Catalyst [a]	*HC*	*H_2O (%)*	*200*	*230*	*300*	*400*	*450*	*500*	*550*
Mn-MFI-106	CH_4	2				31 (23)	39 (36)		46
Ni-MFI-140	CH_4	2				26 (19)	21 (19)		46
Pd-MOR	CH_4	9			87 (44)	9 1 (69)	76 (55)		50
Pd-MFI	CH_4	9			62 (32)	69 (33)			50
In-FER	CH_4	2			60 (14)	58 (32)	56 (30)	41 (27)	107
Ga-MFI-162	CH_4	2					40 (13)		166
Pt-MFI-97	C_2H_4	8.6	17 (19)	10 (10)	4 (3)				169

[a] Numbers next to the zeolite structure designate the exchange percentage of each metal ion.
[b] Numbers in parentheses indicate NO_x conversion at wet condition.

addition, contrary to Pd-exchanged zeolites, little effect of water has been observed by the variation of zeolite structure on the deNO$_x$ activity of Pt-zeolites, suggesting the formation of fine Pt particles on the surface of zeolite. Significant suppression of deNO$_x$ efficiency for CH_4-SCR reaction in the presence of 2% H_2O occurred for an In-MFI catalyst, regardless of the preparation method, although the introduction of second metal ions such as Pt to the In-MFI somewhat improved the extent of the activity loss by water.[107,170]

5.1.5 Supported Noble Metals. Based on the earlier reviews for the HC-SCR reaction with a variety of catalysts,[16,17,19-21,171] Pt, Pd and Rh supported on oxides such as Al_2O_3, SiO_2, TiO_2, and mesoporous material have shown high catalytic activity toward the selective reduction of NO$_x$ by HCs at low temperatures; among them, supported Rh catalysts have revealed the most selective deNO$_x$ activity to N_2.[172] In general, supported Pt appeared to be the most active and robust material for catalyzing a deNO$_x$reaction with HCs; however, the major drawback of these catalysts is the low selectivity toward N_2, producing large quantities of N_2O depending on the supports employed. At least 30 – 50% of NO in feed gas stream is converted to N_2O, with respect to the reaction temperature.[72,172-174] Such a low selectivity can be enhanced by choosing an appropriate catalyst preparation method such as *in situ* sol-gel synthesis of the supports,[76,175] and by using an alternative reductant including toluene.[175a] A narrow bell-shaped activity vs. temperature window was commonly observed over supported noble metal catalysts,[66,74,75,173,176] and their activity was strongly influenced by the supports employed[66,82,177] and the metal dispersion on the surface of the support,[104,173,176,178,179] although the dependence of selectivity on the supports and the metal dispersion was reported to be insignificant.[180,181]

It is difficult to distinguish a net decrease in catalytic activity of supported noble metals by H_2O itself, since most of the earlier works were conducted either in the presence of both H_2O and SO_2 or under only a wet condition.[71,72,76,77,182,183,184,185] Therefore, there are a few limited investigations concerning the effect of H_2O alone on their catalytic activity for HC-SCR reaction;[133,138] however, these studies definitely provide useful information on the water tolerance of the catalyst. 1 wt.% Pt catalyst supported on an MCM-41 (MFO) has shown strong water tolerance during the reduction of NO with C_3H_6 even in the presence of 10% H_2O.[133] This Pt/MFO appeared to be about 8% in activity loss at 250°C (Table 7), and such a wet performance could be recovered to the initial one when feeding the H_2O off. It indicates that the deactivation may be simply due to a competitive adsorption of H_2O and the reactant molecules on the active Pt sites, which was proposed by Schießer *et al.*[186] who observed no alteration of XAS spectra of Pt/MFO exposed to 2.5% H_2O at 500°C for 30 min. A similar result was also observed for 2 wt.% Pt/Al_2O_3[139] and 1.2% Pt/dealuminated Y,[72] as shown in Figure 15. The time-on-stream activity of the former Pt catalyst was quite stable over 700 h even under a wet condition containing 5% H_2O. These supported Pt catalysts can be used for the removal of NO$_x$ from ICEs being operated under LB conditions, particularly for diesel deNO$_x$ applications; however, low selectivity to N_2 formation should be resolved.

Table 7 *Water tolerance of supported noble metal catalysts for the selective reduction of NO_x by HCs*

Catalyst	*HC*	*H_2O (%)*	*Conversion of NO_x (%)*[a] *Reaction temperature (°C)* 250	275	300	350	400	*Ref.*
Pt/MFO	C_3H_6	10	60 (52)					133
Pt/Al_2O_3	C_3H_6	5	44 (44)	62 (55)	52 (40)	30 (24)	24 (11)	138

[a] Numbers in parentheses indicate NO_x conversion at wet condition.

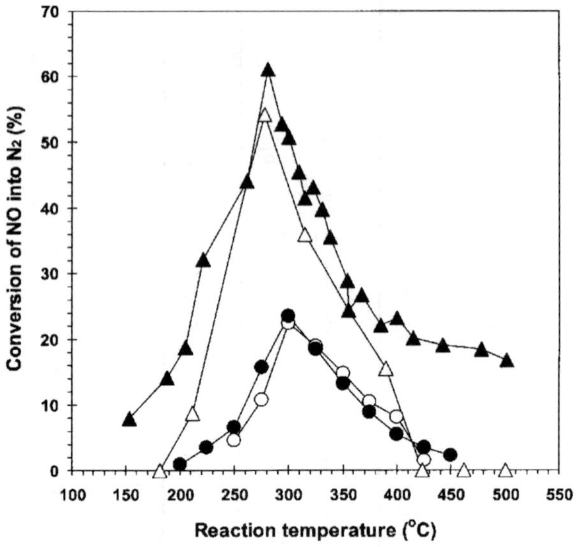

Figure 15 *Effect of H_2O on the NO_x reduction over (▲,△) 2 wt.% Pt/Al_2O_3,[139] and (●,○) 1.2 wt.% Pt/dealuminated Y.[72] Reaction condition: (▲,△) NO 2,000 ppm, C_3H_6 2,000 ppm, O_2 5% and H_2O 0 or 5%; (●,○) NO 1,850 ppm, C_3H_6 300 ppm, C_3H_8 100 ppm, O_2 1%, H_2O 0 or 10%, SO_2 20 ppm and GHSV = 40,000 h^{-1}. Closed and open symbols represent the respective dry and wet $deNO_x$ conversions (adopted from refs. 72 and 139)*

5.1.6 Supported Metal Oxides. Among the supported metal oxide catalysts, Ag catalysts, particularly alumina-supported Ag, have shown strong water tolerance for the present reaction system. Miyadera[85] reported a comparative study on the catalytic activities of Al_2O_3-supported CoO_x, AgO_x, CuO_x, VO_x and CrO_x for the selective reduction of NO with C_3H_6. The 2 wt.% AgO_x catalyst shows high $deNO_x$ activity, *i.e.*, 50 – 80% in the range of temperatures from 350 to 500°C, in dry stream as does 2 wt.% CoO_x/Al_2O_3 (Figure 16 and Table 8). The water tolerance of these Ag and Co catalysts to this reaction also depends on reaction temperatures. Although high-temperature NO conversion is almost maintained even with 10% H_2O, their $deNO_x$ activity decreases by more than 40% at temperatures lower than 400°C. Not only could low NO reduction efficiency be observed for the other three catalysts mentioned above, but their water tolerance was also poor. A similar H_2O effect on the catalytic activity of 2 wt.% AgO/Al_2O_3 catalyst could be observed for NO reduction using the same reductant by Bethke and Kung,[123] as shown in Figure 16. It simply shows that Al_2O_3-supported metal oxides do not possess sufficient water tolerance to use them to control NO_x emission from ICEs. If ethanol as a reductant is employed for NO removal reaction, 2 wt.% AgO_x/Al_2O_3 catalyst,[86] which was used in the initial work by Miyadera,[85] is shown to contain a better $deNO_x$ efficiency with a stronger resistance to water vapor (Figure 16). However, this combination of the catalyst caused significant production of harmful by-products such as CH_3CN,

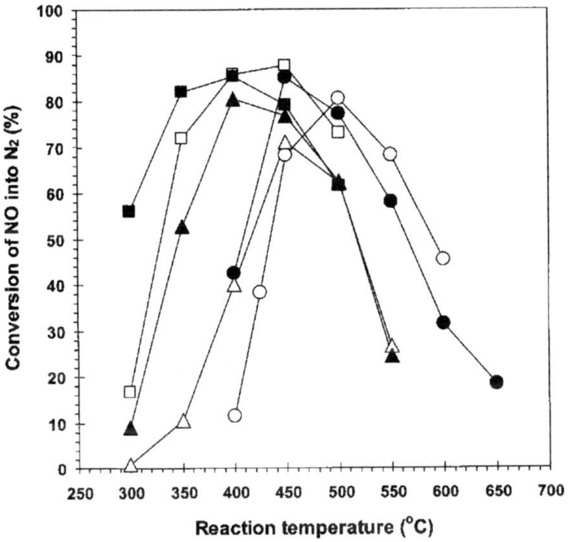

Figure 16 *Effect of H_2O on the NO_x reduction using different reductants over 2 wt.% AgO_x/Al_2O_3. Reaction condition: (▲,△) NO 500 ppm, C_3H_6 500 ppm, O_2 10%, H_2O 0 or 10%, CO_2 10% and GHSV = 6,400 h^{-1};[85] (●,○) NO 1,000 ppm, C_3H_6 1,000 ppm, O_2 6%, H_2O 0 or 1.5% and GHSV = 12,000 h^{-1};[123] (■,□) NO 1,000 ppm, C_2H_5OH 1,250 ppm, O_2 10%, H_2O 0 or 10% and GHSV = 38,400 h^{-1}.[86] Closed and open symbols represent the respective dry and wet $deNO_x$ conversions (adopted from refs. 85, 86 and 123)*

HCN and NH_3. Very recently, such alumina-supported silver catalysts have appeared to be highly active for NO reduction by octanol and octane under a simulated lean-burn composition including 12% H_2O, but no net water tolerance of the catalyst has been documented yet.[187]

A mixture of transition metal oxides and zirconium oxide catalyst prepared by co-precipitation method has been employed for NO reduction by either C_3H_6 or C_3H_8.[188] Cu-Zr-O was particularly active for the reaction, but the $deNO_x$ performance dramatically decreased mainly due to 2.4 % of H_2O contained in the feed gas stream. The water tolerance of alumina-supported InO_x and CoO_x catalysts for NO removal reaction with C_3H_6, has been studied by Maunula *et al.*[91,127] InO_x/Al_2O_3 catalyst in a dry stream reveals 95 % NO conversion at 400°C.[127] If 8% H_2O is subsequently included in the feed gas stream, the catalytic performance notably drops to 55% of NO conversion at the identical reaction temperature, as listed in Table 8. As Mn_3O_4 is physically added to the catalyst (unknown weight ratio), the water tolerance becomes stronger so as to be less than 10% of the decrease in NO conversion within the range of the reaction temperatures covered (Table 8). A similar enhanced water resistance has also been observed for NO reduction with CH_4 over a physical mixture of InO_x/Fe_2O_3 and H-MFI,[130] as listed in Table 8.

The $deNO_x$ activity of CoO_x catalysts in the dry condition significantly varies with respect to the phase of alumina employed as a catalyst support, the Co

Table 8 *Water tolerance of supported metal oxide catalysts for the selective reduction of NO_x by HCs*

Catalyst	*HC*	*H_2O (%)*	*Conversion of NO_x (%)*[a] *Reaction temperature (°C)* 250	300	350	400	450	500	550	600	*Ref.*
Al_2O_3	C_3H_6	10				42 (4)	71 (9)	79 (24)			85
CoO_x/Al_2O_3	C_3H_6	10		13 (3)	76 (5)	76 (20)	60 (42)	43 (44)	15 (22)		85
AgO_x/Al_2O_3	C_3H_6	10		9 (1)	53 (11)	81 (40)	77 (71)	62 (62)	25 (27)		85
CuO_x/Al_2O_3	C_3H_6	10		35 (3)	36 (12)	31 (13)	23 (10)	14 (5)	7 (5)		85
CoO_x/Al_2O_3	C_3H_6	8				88 (16)	81 (49)	62 (66)	23 (54)		91
Cu-saponite	C_3H_6	8				41 (32)	35 (27)	27 (20)			116
Ag-saponite	C_3H_6	8				37 (51)	45 (45)	51 (37)			116
Ag/Al_2O_3	C_3H_6	1.5						43 (34)			123
Cu-pillared clays	C_2H_4	5				55 (32)	71 (50)	81 (76)			124
AuO_x/Al_2O_3	C_3H_6	9.8				30 (42)	57 (72)	45 (52)			125
Mn_2O_3 + Sn-MFI	C_3H_6	5.7		12 (65)	41 (74)	27 (45)		3 (10)			126
InO_x/Al_2O_3	C_3H_6	8			73 (26)	95 (55)	87 (57)	78 (42)	58 (25)		127
InO_x/Al_2O_3 + Mn_3O_4	C_3H_6	8			78 (72)	90 (83)	79 (73)	68 (58)	49 (46)		127
Ga_2O_3/Al_2O_3	CH_4	2.5				6 (0)	13 (2)	28 (7)	70 (21)	62 (29)	128
Cu-aluminate	C_3H_6	10				37 (24)	25 (11)	14 (6)			129
Co-aluminate	C_3H_6	10				47 (28)	55 (47)	54 (44)			129
Ni-aluminate	C_3H_6	10				32 (22)	68 (52)	67 (51)			129
InO_x/Fe_2O_3 + H-MFI	CH_4	3.3				84 (41)	100 (80)	100 (91)	100 (83)	100 (71)	130
Ga_2O_3-Al_2O_3	C_3H_6	9.1		20 (22)	36 (30)	86 (44)	98 (68)	97 (82)	97 (82)	56 (52)	131
CoO_x/Al_2O_3	C_3H_6	20	7 (0)	8 (1)	12 (1)	45 (2)	95 (19)	85 (84)			134
CoO_x/ZrO_2	C_3H_8	8				16 (0)	36 (12)				135
InO_x/CPM[b]	C_2H_4OH	10			0 (10)	0 (10)	13 (18)	25 (39)	36 (71)		137

[a] Numbers in parentheses indicate NO_x conversion at wet condition.
[b] CPM = controlled porosity glass.

precursor and the calcination temperature of the catalyst. The catalyst calcined at 700°C possessed 90 – 60% of dry NO conversion at reaction temperatures ranging from 350 to 550°C. With the addition of 8% H_2O to the dry gas mixture, the NO removal activity below 450°C significantly decreases, while that above 500°C increases, as shown in Table 8. This catalytic behavior of $deNO_x$ catalyst in the presence of water vapor is in good agreement with that of a mechanical mixture of Mn_2O_3 and Sn-MFI (weight ratio of 1:1),[126] Au/Al_2O_3[125] and Ag-saponite[116] on which their wet $deNO_x$ performance within either certain or wide temperature regions has been higher than the dry activities. It is probably associated with the deposited carbonaceous materials that can be readily eliminated when admitting H_2O onto their surface, thereby apparently giving increased catalytic performance, although a promoting role of water vapor for the HC-SCR reaction has been claimed. 3.2% GaO_x/Al_2O_3 catalyst exhibited poor NO conversion of less than 15% within the reaction temperatures covered.[166] The catalytic activity is a strong function of the amount of GaO_x dispersed on the surface of the alumina support. NO conversion of 65% at 500°C was obtained over 35% GaO_x/Al_2O_3 catalyst;[128] however, the catalyst exhibits weak water tolerance (Table 8), although the maintenance of the wet catalytic activity is better than that for Ga-MFI-115. Besides the catalytic systems discussed in the present review, Cu-pillared clays,[124] metal-incorporated aluminates and saponite[116,129] and $Pd/WO_3/ZrO_2$[189] have also been suggested as a $deNO_x$ catalyst with HCs but possess weak water tolerance, as listed in Table 8.

5.2 Hydrothermal Durability of HC-SCR $DeNO_x$ Catalysts. – Among numerous HC-SCR $deNO_x$ catalysts examined, some of the catalysts exhibiting reasonable water tolerance are of particular interest for their commercial use; their practical engine applications subsequently require essentially strong hydrothermal stability of the catalyst, particularly zeolite type catalyst. None of the potential SCR catalysts containing appropriate hydrothermal stability for removing NO_x from mobile sources have been reported yet. This section will concentrate mainly on the time on-stream stability of potential catalysts for HC-SCR reaction.

5.2.1 Fe-Exchanged Zeolites. Overexchanged Fe-MFI catalysts have exhibited extremely high activity and water tolerance for NO reduction with HCs, especially i-C_4H_{10}, as proposed by Feng and Hall[52,53] in the early section. Thus, these SCR catalysts are anticipated to be among the most promising catalyst candidates for controlling NO_x from mobile sources. Adequate hydrothermal stability, commonly recognized as one of weak points for zeolite, is required for this purpose. Figure 17 represents a 2500-h continuous operation result of overexchanged Fe-MFI-183 for the reduction in the presence of 20% H_2O as well as of 150 ppm SO_2. This catalyst appeared to have approximately 95% stable NO_x conversion activity for 5 h with a clean NO-*iso*-C_4H_{10}-O_2 reaction. Such stability in $deNO_x$ performance was consistently maintained during 116 h of the on-stream time even in the presence of 20% H_2O. After this time on-stream operation, 150 ppm SO_2 was subsequently introduced to the wet reaction com-

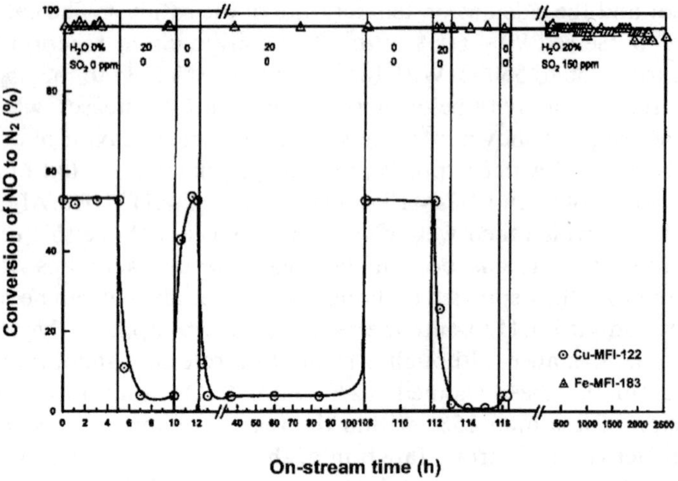

Figure 17 *Effect of H_2O and SO_2 on the conversions at 500°C over Fe-MFI-183 vs. Cu-MFI-122 compared as a function of time on stream. The weight of catalyst was 50 mg in each case and the standard feed gas composition was: NO, 2,000 ppm; i-C_4H_{10}, 2,000 ppm; O_2, 3%. H_2O vapor (20%) and SO_2 (150 ppm) were added in separate metered feed streams and were substituted for a portion of the He carrying gas. The calculated GHSV was 42,000 h^{-1} (adopted from ref. 53)*

position, but there was no alteration of the catalytic activity even for 2,500 h. This behavior is quite a contrast to Cu-MFI-122 catalyst, revealing a completely reversible deactivation phenomenon even during the cyclic feed of H_2O into the feed gas stream. Even after a hydrothermal excursion for 2,500-h time on-stream, the Fe-MFI-183 catalyst showed the same ^{27}Al MAS NMR spectrum as a fresh sample, as compared in Figure 18. The reason why this Fe-MFI catalyst possesses such high durability under the wet condition has not been clarified yet in the literature; however, it seems probably due to the absence of Brønsted acid sites on the catalyst surface after exposing to the wet mixture for 2,500 h, confirmed by ^{1}H MAS NMR spectrum of the Fe-MFI-183.[53]

Unfortunately, the preparation method proposed by Feng and Hall[52,53] has been irreproducible and quite sensitive to the origin of the parent zeolite and its preparation history. A chemical vapor deposition technique may provide reasonable reproducibility for obtaining overexchanged Fe-MFI catalysts even with the parent zeolites from different manufacturers.[54,55] The overexchanged Fe-MFI-300 catalyst prepared *via* the CVD route appeared to contain excellent hydrothermal stability with high deNO_x reduction activity, as shown in Figure 19. Approximately 75% of the initial NO conversion efficiency at 375°C under the dry feed was maintained for 16-h time on-stream and then was not altered when exposed to 10% H_2O for 30 h. Subsequent cyclic exposure of the catalyst to the water vapor exhibits practically negligible performance loss in this reaction. It implies that the CVD preparation technique can provide active and durable Fe-MFI catalyst even under high water vapor pressure. It may be again anticipated that the absence of the protonic sites in the framework of zeolites is

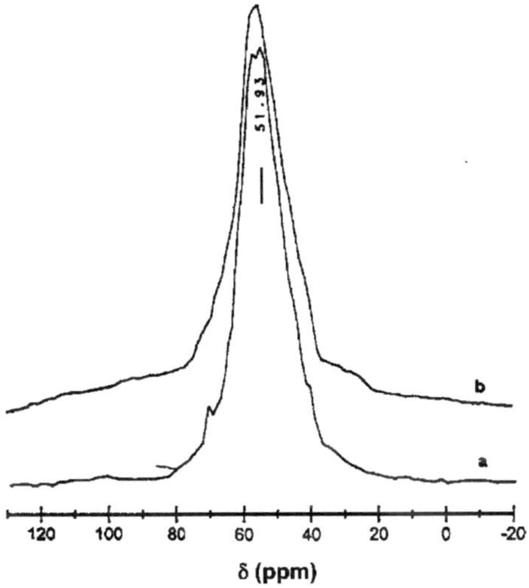

Figure 18 *(a) 27Al MAS NMR spectrum of freshly prepared Fe-MFI-183 compared with (b) the spectrum after the 2,500-h test. The integrated intensities for (a) and (b) were 0.7 and 0.6 mmol Al/g, respectively*[53]

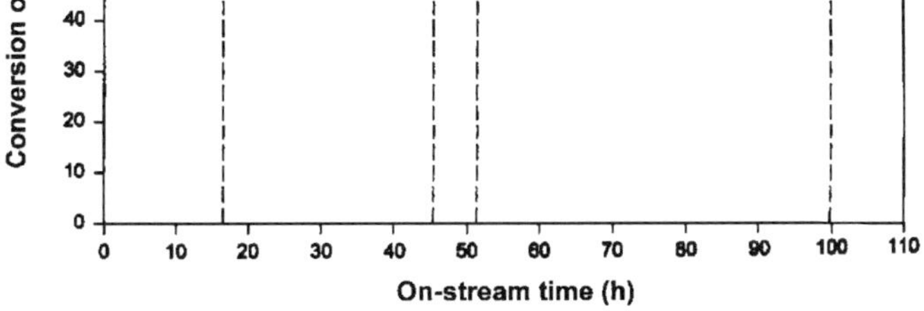

Figure 19 *Effect of the introduction of H_2O on the conversion of NO to N_2 at 375°C over Fe-MFI-300. Reaction condition: NO 2,000 ppm, i-C_4H_{10} 2,000 ppm, O_2 3%, H_2O 0 or 10% and GHSV = 42,000 h^{-1} (adopted from ref. 55)*

probably responsible for high hydrothermal stability, as reflected from the ion exchange level of the catalyst.

Sachtler and coworkers[55] have proposed a plausible reason why the overexchanged Fe-MFI catalysts can possess high $deNO_x$ activity and strong hydrothermal durability. The presence of oxo-, hydroxo- and oxygen-bridged Fe ions as much as possible is crucial for such high catalytic properties. As already observed, such species in Cu-MFI catalysts appear to increase on the catalyst surface as the copper exchange level increases.[190,191] Thus, the CVD method consisting mainly of subliming $FeCl_3$ onto an H-MFI may easily allow that circumstance in the zeolite infrastructure by the elimination of all the acid sites, as indicated by IR spectra of the overexchanged Fe-MFI-300 on which no bands are observed at 3750 and 3610 cm^{-1} due to the respective isolated silanol and acidic bridged-hydroxyl groups. Subsequent water washing after the sublimation may also produce an oxygen-bridged binuclear Fe complex on the surface of the catalyst, such as $[(HO)Fe\text{-}O\text{-}Fe(OH)]^{2+}$, in the zeolite cavities, which could be supported by TPR profiles of H_2 and CO and their consumption ratio to Fe. *In situ* XAS studies of Battiston *et al.*[139] for an Fe-MFI have represented the existence of binuclear Fe-O-Fe species which are catalytically active for NO reduction with *iso*-butane. These binuclear complexes, when exposed to a flowing mixture of *iso*-butane in He, are reduced to Fe-□-Fe vacant sites; however, exposure of the reduced sample to an NO-*iso*-C_4H_{10}-O_2 mixture results in the oxidation of vacancies to the Fe-O-Fe complexes. *In situ* ESR investigations for an Fe-MFI-210, prepared by using the sublimation method, exhibited no chemisorption of H_2O on Fe^{3+} ions in the form of tetrahedral and distorted tetrahedral at 200°C.[192] Characterization of an Fe-MFI catalyst by acquiring IR spectra after exposure of NO, C_3H_8, and O_2 onto activated and wet surfaces indicated the formation of a surface C-H-N-O compound which can be partly decomposed into NCO species at temperatures higher than 250°C.[192a] This process is not affected by the presence of H_2O vapor. Consequently, it is probable that binuclear oxygen-bridged Fe ions are quite active for NO removal reaction with HCs, particularly *iso*-butane, and possess highly durable catalytic properties even under hydrothermal conditions. It should be noted that the formation of binuclear Fe on the surface of Fe-MFI prepared by CVD is still controversial.

5.2.2 Cu-Exchanged Zeolites. Few investigations concerning the hydrothermal stability of these $deNO_x$ catalysts for NO removal reaction with HCs can be found in the literature. For Cu-SAPO-75 and Cu-MFI aged under flowing of 3% H_2O in air for 2 h at high temperatures such as 500, 700 and 800°C, the Cu-SAPO even after a treatment at the former two temperatures has revealed strong hydrothermal stability, but aging at 800°C caused significant deactivation of both catalysts.[193] Such a dependence of the hydrothermal stability on aging temperatures is also observed for Ce-Ag-MFI catalyst containing different ratios of Ce/Al and Ag/Al.[194] The hydrothermal durability of Cu-zeolites also depended on Si/Al ratio of the parent zeolites.[195] Cu-MFI catalysts with high Si/Al ratio, if the Cu content is similar or comparable to each other, exhibit better activity maintenance, indicating that the hydrothermal stability can be improved by the increase of the Si/Al ratio of the catalyst.

Chung *et al.*[11] have attempted to illustrate the major role of Si/Al ratio for improving the hydrothermal stability of synthetic Cu-MOR catalysts containing a variety of Si/Al ratios, as listed in Table 9. Catalysts containing Si/Al ratios greater than the parent zeolite were prepared by dealumination in an aqueous HCl solution. As shown in Figure 20, zero activity was observed at temperatures

Table 9 *Physicochemical properties of MOR structure zeolite catalysts with and without a hydrothermal aging*

				Surface area (m^2/g)	
Catalyst[a]	Cu content (wt.%)[b]	Si/Al[b]	Cu/Al[b]	Fresh	Hydrothermally-aged[c]
H-MOR (5)		5		449	
H-MOR (10)		10			
H-MOR (20)		20		449	
Cu-MOR-32 (6)	2.02	6	0.16	368	22
Cu-MOR-60 (5)	4.20	5	0.30		
Cu-MOR-68 (12)	2.55	12	0.34	434	154
Cu-MOR-28 (12)	1.03	12	0.14		
Cu-MOR-90 (22)	1.73	22	0.45	450	330
Cu-MOR[d]-50 (4)	4.37	4	0.25	179	
Cu-MOR[d]-48 (10)	1.84	10	0.24	232	26
Cu-MOR[d]-56 (14)	1.75	14	0.28		
Cu-MOR[d]-62 (19)	1.64	19	0.31	128	
Cu-MFI-162 (26)	2.90	26	0.81	344	249 27[e]

Note. Adopted from Tables 1 and 2 in ref. 11.
[a] Numbers next to the zeolite structure designate the exchange percentage of Cu ions.
[b] Based on fresh samples.
[c] Under flowing 10% H_2O in He at 800°C for 24 h.
[d] Natural zeolite consisting mainly of a MOR structure.
[e] At 900°C.

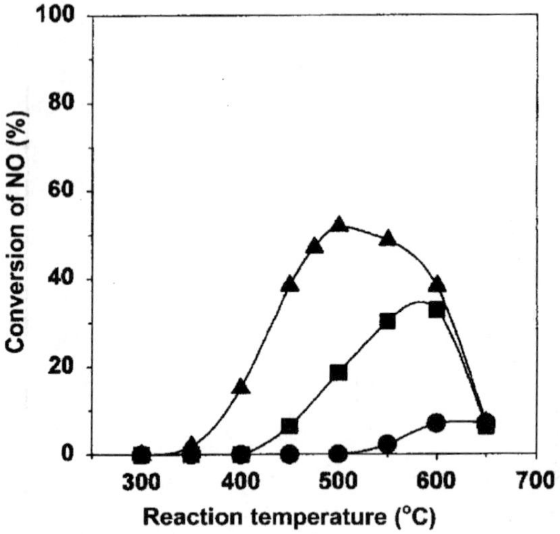

Figure 20 *Hydrothermal stability of dealuminated synthetic Cu-MOR catalysts: (●) Cu-MOR-32 (6); (■) Cu-MOR-68 (12); (▲) Cu-MOR-90 (22). Reaction condition: NO 1,200 ppm, C_3H_6 1,600 ppm, O_2 3.2 %, H_2O 10%, CO 3,000 ppm, H_2 1,000 ppm and CO_2 10%*[11]

below 550°C under simulated lean NO_x condition for Cu-MOR-32 (6) catalyst without dealumination and hydrothermally aged at 800°C in a flowing mixture of 10% H_2O/90% He for 24 h. The dealuminated Cu-MOR catalysts exhibit strong activity maintenance with respect to the Si/Al ratio of the catalysts. For example, NO conversions of 50, 20 and 0% at 500°C are achieved for the respective aged Cu-MOR-90 (22), Cu-MOR-68 (12) and Cu-MOR-32 (6) catalysts. After the hydrothermal aging, Cu-MOR-60 (5) catalyst in Table 9 revealed almost identical deNO_x activity to that of Cu-MOR-32 (6).[11] It clearly shows that the hydrothermal stability can be beneficially modified by the catalyst dealumination, thereby increasing the Si/Al ratio of the catalyst. The dependence of hydrothermal stability on the Si/Al ratio for MOR type natural zeolite-based Cu catalyst essentially exhibited the similar trend to the synthetic Cu-MOR. This may be a typical example for the close relationship of the Si/Al ratio of zeolites with the hydrothermal durability of the catalyst. The characterization of these aged Cu-MOR catalysts by acquiring BET, XRD and ESR measurements revealed significant destruction of their structure, thereby resulting in partial sintering of active Cu^{2+} ions to CuO species on the catalyst surface. This may be the reason why the aged Cu-MOR catalysts still exhibited lower deNO_x activity compared to the fresh ones. Such a collapse of zeolite structure upon the hydrothermal aging could typically be observed by BET surface area measurements, as listed in Table 9. Consequently, dealumination can offer better hydrothermal stability of zeolite-based SCR catalysts, but this approach does not seem to be advisable because of the corresponding loss of exchangeable sites in the zeolites.

Copper ions in zeolites are mainly responsible for NO_x reduction activity; therefore, deactivation would occur if alteration of their state or the zeolite structure were accompanied. The origin of the deactivation of Cu-exchanged zeolites due to their hydrothermal excursion has been published in several literatures. Solid state ^{27}Al MAS NMR spectra of a Cu-MFI catalyst appeared to have a loss of 23% of framework aluminum after hydrothermal aging at 410°C in flowing 10% H_2O/90% air for 113 h.[196] Kharas *et al.*[111] reported the formation of CuO crystallites on the surface of Cu-MFI-387 exposed to simulated lean NO_x condition at 800°C for 1 h, which is in good agreement with other results acquired for Cu-MFI-678 and Cu-MOR catalysts.[11,197] On the other hand, XRD result by Matsumoto *et al.*[198] presented that there is no detection of the formation of CuO crystallites for a Cu-MFI-110 catalyst aged at 800°C in a simulated lean NO_x mixture. Spectroscopic characterization of fresh and deactivated Cu-MFI-113 catalysts by ESR, XRD and FTIR in addition to BET surface area measurements and H_2-TPR revealed no appreciable change in the zeolite framework of the deactivated catalyst which had lost almost 50% of the original conversion activity after wet operation at 400°C for 100 h in a mixture of NO, C_3H_8 and O_2 with 10% H_2O.[199] Three major Cu species such as isolated Cu^{2+}ions, $[Cu\text{-}O\text{-}Cu]^{2+}$oxocations and CuO crystallites were detected in the fresh Cu-MFI-113, while Cu ions highly-dispersed in Al_2O_3 and a $CuAl_2O_4$ were subsequently present on the surface of the deactivated Cu-MFI catalyst as verified by electron paramagnetic resonance (EPR) technique (Figure 21) and TPR. It

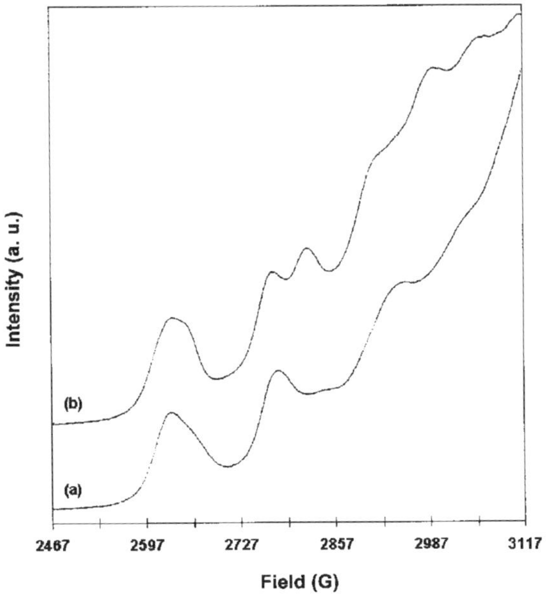

Figure 21 *EPR spectra of copper hyperfine structures of (a) fresh Cu-MFI-113 and (b) deactivated Cu-MFI-113 after treatment under wet catalysis condition for 44 h (adopted from ref. 199)*

indicates that such a hydrothermal excursion induces the elimination of the framework aluminum, and the fine Al_2O_3 clusters interact with the small CuO particles to form the subsequent Cu species which are much less active for NO_x reduction. Grinsted *et al.*[196] proposed the extent of the loss of framework aluminum, of Cu-MFI catalyst, almost equal to the amount of protonic sites on the surface of fresh catalyst. Therefore, it suggests that the complete removal of all protonic sites in Cu-only exchanged zeolites is crucial to obtain their stronger hydrothermal durability for wet deNO_x catalysis. It would be of particular interest to prepare Cu-exchanged zeolites by subliming $CuCl_2$ and testing them for NO_x reduction reaction to compare with the conventional ion exchange technique, as in the case of preparing overexchanged Fe- and Co-zeolite catalysts.

5.2.3 Co-Exchanged Zeolites. Hydrothermal durability of Co-zeolites usually depends on the nature of the parent zeolite, Co exchange level, preparation method, etc. Existence of both Co and Brønsted acid sites in zeolites can play a synergistic role for catalyzing NO_x reduction reaction with HCs; however, the protonic sites induce catalyst deactivation by H_2O.[46,113,151] Not only can the activity of Co-zeolites be enhanced when containing high Co exchange levels, but their durability also improves.[51,96,98,99,146,152] Pieterse *et al.*[113] have reported that Co-MFI-80 catalyst obtained by wet ion exchange lost its initial NO_x conversion efficiency, by 25%, after CH_4-SCR reaction with 5% H_2O at 450°C for 35 h, which was quite similar to the time-on stream trend of the deNO_x activity over Co-impregnated H-MFI catalyst. Any IR bands due to OH stretching did not appear on the surface of the catalyst, indicating all the cations are still retained as cation exchange sites on the surface of the zeolite even after the hydrothermal aging. It implies that the catalyst preparation route may be one of the critical factors related to the hydrothermal durability of Co-zeolites for HC-SCR reaction.

With C_3H_8 as a main reductant for NO_x reduction in a simulated exhaust gas stream containing 9% H_2O, a distinctive wet activity maintenance with respect to the time on-stream has been exhibited for Co exchanged on a variety of zeolite structures including BEA, MFI, MOR, FER, and FAU.[144,158,200] Not only could the highest deNO_x activity be shown for Co-BEA catalysts consisting of different ratios of Si/Al and Co exchange, but the strongest hydrothermal durability was also attained. An optimal exchanging content of Co exists to offer the preferable performance of those catalysts for the present HC-SCR reaction and a low ratio of Si/Al in zeolite is necessary to maximize the conversions of NO_x and C_3H_8. 80% of NO_x conversion activity was achieved over Co-BEA-98 catalyst when 0.3 ppm SO_2 was subsequently present. Contrary to Co-MFI-106, this catalyst was slightly deactivated by 10% of the initial efficiency even for *ca.* 4,000 h, as compared in Figure 23. Relatively poor durability for the same reaction is observed over a Co-MOR-74 catalyst.[202] This indicates that the hydrothermal stability of Co-zeolites depends significantly on the structure of zeolite supports. It should be noted that ferrierite type zeolites might be a better support for NO removal reaction with CH_4 even in the presence of water vapor.[201]

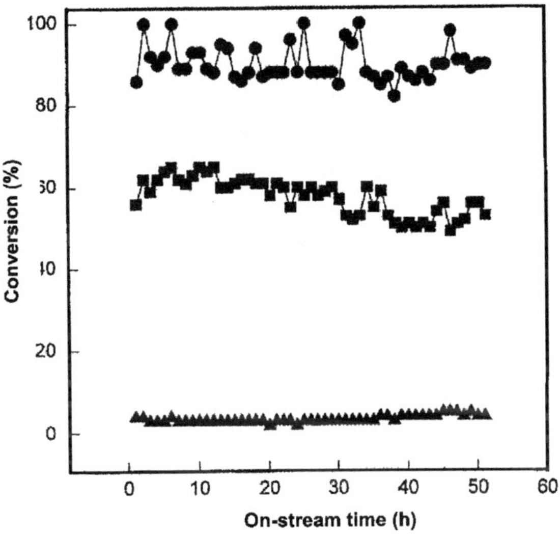

Figure 22 *Hydrothermal durability of Co-MFI-226 prepared by sublimation technique. (●) Conversion of NO into N_2; (■) conversion of i-C_4H_{10} into CO_2; (▲) conversion of i-C_4H_{10} into CO. Reaction condition: NO 2,000 ppm, i-C_4H_{10} 2,000 ppm, O_2 3 %, H_2O 10%, GHSV = 42,000 h^{-1} and T = 400°C*[154]

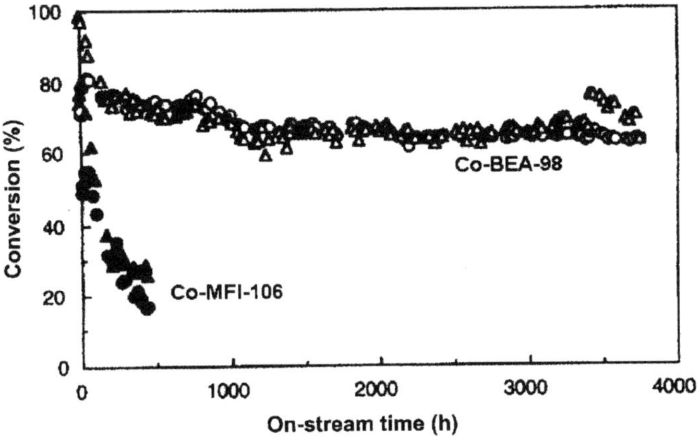

Figure 23 *Durability of activity of HC-SCR by C_3H_8 on Co-BEA-98 and Co-MFI-106. Reaction condition: NO = 150 ppm, C_3H_8 = 500 ppm, O_2 = 10%, H_2O = 9%, SO_2 = 0.3 ppm, CO_2 = 6%, CH_4 = 1000 ppm, CO = 500 ppm, H_2 = 250 ppm, He balance, T = 400°C, GHSV = 15,000 h^{-1}. (●,○) NO_x conversion; (▲,△) C_3H_8 conversion). Si/Al ratios of the Co-BEA and Co-MFI are 16.3 and 50, respectively*[200]

Howe and coworkers[203] have examined the effect of hydrothermal treatment at 800°C for 24 h in 15% H_2O in air on the performance of Co-MFI and observed the elimination of the framework Al of zeolite, causing loss of cation exchange capacity and residual Brønsted acidity with a concomitant decrease in the catalytic activity for CH_4-SCR reaction. Pre-exchange of the zeolite with La^{3+} ions could stabilize such a hydrothermal dealumination process, thereby significantly preventing impairment of the catalyst performance in the wet condition. Prior introduction of K^+ or Cs^+ ions likewise into Co-TSZ zeolites increased their durability during treatment with a simulated model gas with 3% H_2O for 15 h at 800°C.[204] This indicates that selecting suitable second metal ions can assist in overcoming the weak hydrothermal durability of Co-zeolites. Simultaneous use of noble metals, representatively Pd and Pt, with Co-exchanged zeolites appears to be an advisable approach to prevent or at least reduce their deactivation during NO_x removal reaction with HCs in the presence of H_2O; however, the influence of such approaches in catalyst design on the time on-stream stability of those catalysts under hydrothermal condition has been unknown.[99,148,149]

5.2.4 Other HC-SCR Catalysts. Pd-exchanged zeolites are widely reported to exhibit poor stability even in a dry gas stream. The durability of Pd-zeolites for HC-SCR reaction with and without water vapor has been predominantly affected by the Si/Al ratio of parent zeolites, Pd exchange level, and the zeolite structure employed. Pd-MFI-14 and Pd-silicallite-40, possessing the respective Si/Al ratios of 25 and 131, were aged at 800°C for 6 h in flowing N_2 with 10% H_2O to examine their hydrothermal stability for CH_4-SCR reaction and both catalysts were completely deactivated even in a dry stream. This is due to the agglomeration of active Pd species to PdO crystallites on which CH_4 combustion is generally preferable, as characterized by XRD, TEM and NMR.[161] With a Pd-MFI catalyst containing low Si/Al ratio, NO_x conversion efficiency was higher and relatively stable even in wet feed gas stream, while Pd-MFI catalysts with higher Si/Al ratio showed significant impairment of their hydrothermal durability.[106] Similarly, Raman spectra of the deactivated catalysts produced a 648-cm^{-1} intense band due to PdO formation, suggesting well dispersed Pd is the active center for the present reaction system. Hydrothermal durability of Pd-zeolites also depended on the original nature of zeolite structure as shown in Figure 24.[50] The Palladium ingredient, exchanged in a variety of zeolite structures containing comparable exchange level, exhibits the distinctive maintenance of the time on-stream activity for CH_4-SCR reaction with 9% H_2O. Note that dry deNO_x activity at 450°C of the Pd-MOR-47 does not decrease during a 30-h continuous operating time but does gradually for the Pd-MFI-58 catalyst. Recently, Ogura *et al.*[119] again reported such dependence.

Few, if any, time on-stream performance data for HC-SCR reaction have been reported over supported Pt such as Pt-zeolites and Pt/solid acids. Iwamoto and coworkers[205] observed small loss of initial activity of a Pt-MFI catalyst during an on-stream period of 200 h in a model exhaust gas stream containing 10% H_2O after which it was no longer deactivated even for 800 h. Similar results were recently reported for 1 wt. % Pt/MFO on which less than 10% of NO conversion

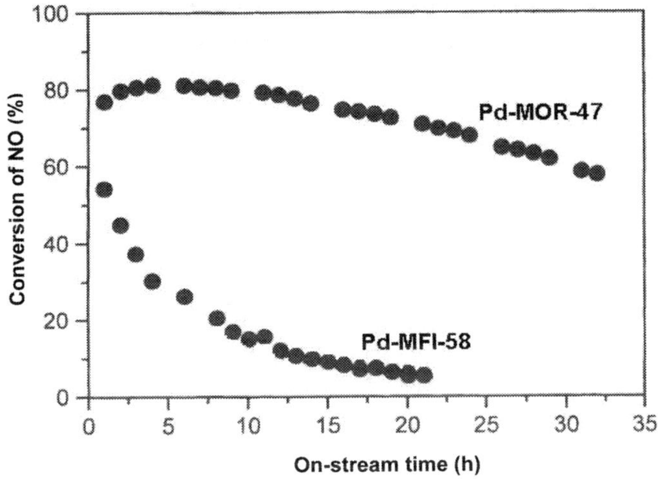

Figure 24 *Hydrothermal durability of Pd-exchanged zeolites in the selective reduction of NO_x by CH_4. Reaction conditions: NO 150 ppm, CH_4 2400 ppm, O_2 10%, H_2O 9%, GHSV = 15,000 h^{-1} and T = 450°C (adopted from ref. 50)*

is suppressed by the presence of 10% H_2O and 500 ppm SO_2.[133] The catalyst durability of supported and bare metal oxides has been observed as a function of time on-stream in the presence of H_2O.[206-209] A 2 wt.% CoO_x/Al_2O_3 catalyst, prepared by impregnating a specially-synthesized Al_2O_3 with a $Co(NO_3)_2$ solution following calcinations at 800°C, reveals no decrease in initial NO_x conversion activity of *ca.* 45% at 550°C for 124 h in a model gas stream with 5% H_2O.[206] Such strong durability was also observed for a 10 wt.% CoO_x supported on montmorillonite which, after 120 h of time on-stream, lost about 15% of its original activity for CH_4-SCR reaction with 2% H_2O, although cobalt oxides on other supports such as MgO, SiO_2 and Al_2O_3 were deactivated by more than 30% within 30 h.[208] High durability of a AgO_x-Al_2O_3 obtained by a single-step cogelation technique was recently reported by Keshavaraja *et al.*[210] Initial deNO_x conversion of this catalyst irreversibly fell down by 30% after *ca.* 15 h of continuous CH_4-SCR reaction at 550°C with 10% H_2O and 30 ppm SO_2, which was predominantly associated with the water vapor contained in the feed gas stream.

5.3 Beneficial Modification of HC-SCR DeNO_x Catalysts to Improve Hydrothermal Stability. – Impairment of the hydrothermal durability of HC-SCR catalysts, particularly metal-exchanged zeolites which are of interest in high-temperature deNO_x applications, can be either prevented or at least reduced by selecting appropriate parent zeolite structure, employing suitable preparation technique, modifying physicochemical properties, and adding subsequent cation. The former two approaches have been extensively described in the previous section and the other ones will be mainly discussed.

Prior to introducing Cu ions into natural MOR (NMOR), its dealumination, thereby increasing Si/Al ratio, has allowed less deactivation of Cu-NMOR, which had been aged at 800°C for 24 h in a flow of 10% H_2O in He, for NO reduction in a wet model lean-burn exhaust stream.[11,211] Three samples of Cu-NMOR containing comparable Cu/Al ratios revealed distinctive NO reduction activity, as shown in Figure 25. The hydrothermal aging causes the complete loss of the NO_x conversion activity of the Cu-NMOR possessing the lowest Si/Al ratio, while peak deNO_x efficiencies greater than 20% remained for the Cu-NMOR catalysts with higher Si/Al ratios even after the dealumination process. This clearly represents that the hydrothermal stability is a strong function of Si/Al ratio, which is in good agreement with an earlier study on the deNO_x activity of Cu-MFI-80 (27) and (11).[212] Similar results for CH_4-SCR reaction in the feed gas stream including 9% H_2O have been observed over Pd-MFI-10 catalysts containing distinctive Si/Al ratios of 15 and 25.[106] The surface hydrophobicity of the catalyst is associated with the ratio of Si vs. Al in the framework structure of zeolite,[213] anticipating that the surface of zeolite-based catalysts becomes hydrophobic as the Si/Al ratio of zeolite increases. Therefore, favorably adjusting Si/Al ratio of parent zeolites may be one useful approach to improving the hydrothermal durability of metal-exchanged zeolite catalysts.

Second ingredients including Co, Ca, Cr, Sr, Ag, Ce, Ba, Mg, Mo, Fe, Sn and so forth can be simultaneously employed to improve hydrothermal durability of SCR catalysts such as metal-exchanged zeolites, supported metals and their

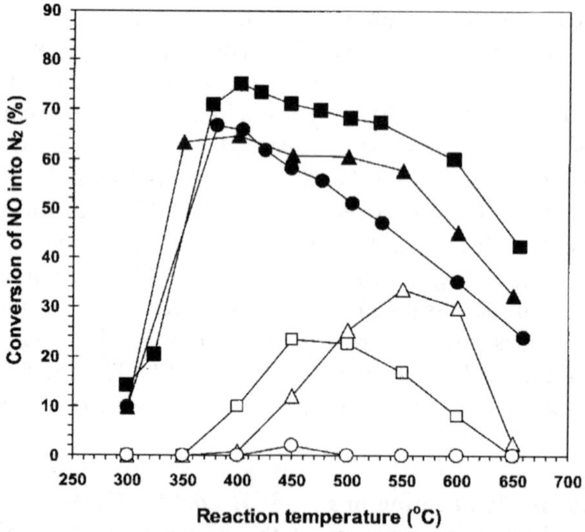

Figure 25 *Effect of Si/Al ratio on hydrothermal stability of dealuminated natural MORs: (●,○) Cu-MOR-48 (10); (■,□) Cu-MOR-56 (14); (▲,△) Cu-MOR-62 (19). Reaction condition: NO 1,200 ppm, C_3H_6 1,600 ppm, O_2 3.2 %, H_2O 10%, CO 3,000 ppm, H_2 1,000 ppm and CO_2 10%. Closed and open symbols represent the respective catalysts unexposed and exposed to 10% H_2O at 800°C for 24 h (adopted from refs. 11 and 211)*

oxides, and bare solid materials, although those components are subsequently introduced to mainly modulate selectivity to N_2O formation, and thermal stability of active constituents and their supports. When *ca.* 2.2% Co was loaded onto Pd-exchanged zeolites including BEA, MFI, FER and MOR, by using incipient wetness, the Pd-Co-FER and -MOR catalysts showed stable time-on-stream activity for CH_4-SCR reaction at 450°C for 50 h in the presence of 5% H_2O.[147] The Co- and Pd-only catalysts revealed poor hydrothermal durability for this reaction system.[113] These results simply show the significant role of Co for stabilizing Pd species during wet operation at high temperatures.[119] Concomitant use of Ce with Ag-MFI,[194] La with Co-MFI[203] and Ag, Cr, and La with Cu-MFI[214-216] has been recognized to add stronger catalytic durability for NO removal reaction by HCs in a stream with significant amounts of H_2O than single metal-exchanged zeolites. It indicates that HC-SCR catalysts can be modified through subsequently selecting suitable elements to allow better hydrothermal durability for wet deNO_x catalysis. In addition, the introduction of small amounts of Na and Ba to Cu-based MOR catalyst enhances the catalytic performance for NO reduction by C_3H_6 in the wet condition.[195,212] A similar role of second metals has been documented for Pt-In-FER,[107] Pt-Co-MFI,[148] Pt-Co-MOR[149] and Pd/Ag/H-MOR[217] for wet HC-SCR deNO_x catalysis. However, whether or not the guest components play a beneficial role for improving the hydrothermal stability has hardly been known for all these catalysts discussed.

5.4 Cause of the Deactivation of HC-SCR DeNO_x Catalysts by H_2O. – Some of the HC-SCR DeNO_x catalysts developed so far possess reasonable water tolerance and hydrothermal durability, in addition to high catalytic activity even under the wet condition containing significant amounts of H_2O vapor, and another group of the SCR catalysts experience severe deactivation in their catalytic activity in the presence of H_2O, as summarized in Tables 3-8. Overexchanged Fe-zeolites, supported Pt and Ag and InO_x-based metal oxides can be listed in the first category, and the latter one includes Cu-zeolites as well as other promising catalysts. Catalytic properties, particularly water tolerance and hydrothermal stability, for both groups of the catalysts depend predominantly on feed gas composition, reaction temperature, water content, support and its modification, guest ingredient, and preparation technique employed for the catalytic reaction system. Of the relatively huge bodies of HC-SCR catalytic system studies to date, little has been understood, particularly on the cause of the deactivation of SCR catalysts during wet deNO_x catalysis. In this section, two distinctive catalytic systems, Fe-exchanged zeolites possessing high activity and durability even in the existence of H_2O and SO_2, and Cu-exchanged zeolites revealing high activity but poor water tolerance and hydrothermal stability, will be particularly covered to induce the guidelines for the development of 'better' deNO_x catalyst by HC for future Advanced ICEs.

5.4.1 Fe-Exchanged Zeolites. Since overexchanged Fe-MFI containing a ratio of Fe/Al ≈ 1 appeared to exhibit extremely high catalytic activity and strong hydrothermal durability for the selective reduction of NO_x with *iso*-butane as invented by Feng and Hall,[52,53] this has been the most promising catalyst for deNO_x application to ICEs; however, their preparation technique can hardly be reproducible, as discussed. Instead, chemical vapor deposition *via* subliming $FeCl_3$ onto the porous surface of a H-MFI following subsequent washing and calcination was shown to be repeatable for obtaining overexchanged Fe-MFI catalysts, although these possessed somewhat lower and very narrow deNO_x activity window. But strong durability was still maintained under high-temperature wet conditions, as first developed by Sachtler and coworkers.[55] However, the nature of the active species of these overexchanged catalysts and the origin of their extremely-strong hydrothermal stability has hardly been elucidated yet. As already stated in the present review, all of the protonic sites on the surface of the catalyst offering ion exchange positions in the infrastructure of zeolites may need to be completely replaced by Lewis acidic ones to obtain stronger water tolerance and their resistance to hydrothermal excursion, although the protonic sites also play a role for reducing NO_x with HCs. With the catalyst preparation technique proposed by Feng and Hall,[52,53] iron species in the ferrous oxalate solution mainly exist in the form of $Fe(OH)^+$but not Fe^{2+} ions which can be readily oxidized to trivalent Fe, thereby precipitating FeOOH in the zeolite cavities, as proposed earlier.[218] These circumstances make it possible to allow the exchange of all Na^+cations to $Fe(OH)^+$and there is no presence of any Brønsted acid sites, which are assumed to be responsible for the hydrothermal destruction of the catalyst. An oxygen-bridged binuclear Fe complex and clusters of iron and oxygen atoms have also been suggested to be plausible active species for NO removal reaction by HCs, particularly i-C_4H_{10}.[55,219]

An overexchanged Fe-MFI catalyst prepared by the CVD route was characterized by acquiring FTIR, HRAEM and TPD.[55] All the Brønsted sites could be completely removed upon subliming $FeCl_3$, probably by exchange with $(Fe_2Cl_4)^{2+}$entities *via* the following reaction.

$$Fe_2Cl_6 + 2H^+ \rightarrow [Fe_2Cl_4]^{2+} + 2HCl \qquad (2)$$

Occurrence of this probable reaction during CVD was supported by FTIR spectra of the catalysts, as shown in Figure 26. Peaks at 3750 and 3610 cm^{-1} due to the respective isolated silanol and acidic hydroxyl groups disappear after the sublimation process. Subsequent washing by using distilled deionized water following calcination can cause their restoration, suggesting a mobility of such iron species, thereby giving clustered moieties. After that treatment, the Fe-O-Fe dimer species such as binuclear $[OH\text{-}Fe\text{-}O\text{-}Fe\text{-}OH]^{2+}$ is formed as a major active site responsible for the high SCR activity and its strong hydrothermal durability for reducing NO_x, which is in good agreement with the similar structural observation *via* XAS (XANES and EXAFS) studies.[139,220]

Battiston *et al.*[221] have proposed a structure of binuclear Fe species based on XAS characterizations of an Fe-MFI-300 catalyst that underwent different heat treatments. Washing of the catalyst results in the structure (a) and its exposure to room temperature in a flow of pure He leads to the structure (b), as shown in Figure 27. The Fe-O2 shell near 1.97 – 1.99Å could be assigned to two oxygen atoms in the tetrahedra of the zeolite. The Fe-O3 shell at 2.02-2.09Å is due to an

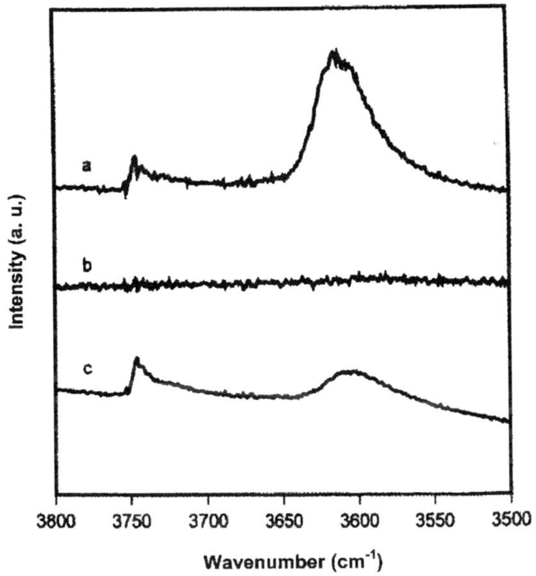

Figure 26 *FTIR spectra in the OH stretching region of (a) pure H-MFI, (b) after sublimation, and (c) after calcination*[55]

oxygen atom of the zeolite and to an adsorbed water molecule, respectively. Exposure of the catalyst to a flow of either He or its mixture with gas-phase oxygen at moderate temperatures less than 150°C causes the desorption of the water on Fe sites (structure (c)), and further heating to 350°C in an He flow removes the bridged oxygen atom (O1) in the Fe-O-Fe to form the structure (d) which can readily return to the structure (c) in the presence of gas-phase excess oxygen. The proposed structure (e), appearing at 350°C in the presence of excess oxygen in which typical HC-SCR reactions occur, is quite close to the binuclear iron complex proposed earlier.[55] The high reactivity of the bridging oxygen in the closest Fe-O shell may be responsible for high $deNO_x$ activity of Fe-MFI catalysts; and their strong water tolerance and hydrothermal durability are probably associated with the weak interaction of the Fe sites with water molecules as anticipated by the structure (c). This interpretation may be also supported by *in situ* ESR measurements of Kucherov *et al.*[192] who reported no change in the ESR spectra of Fe-MFI-210 which had undergone wet and dry cycles in the range of temperatures from 200 to 500°C (Figure 28). Furthermore, the bonding of water molecules with the formation of coordinatively unsaturated zeolite structures at 20°C has been quite weak. It indicates that water molecules are difficult to chemisorb on this catalyst even at 200°C, which is contrary to Cu-MFI on which Cu^{2+} ESR spectra have been altered in the presence of water vapor even at 400°C.[222] It is proposed that substantial investigations are required, including H_2O-TPD and molecular adsorption on wet Fe-MFI surface combined with *in situ* DRIFTS technique. Based upon the literature searched so far, overexchanged Fe-MFI catalysts after proper activation processes are quite

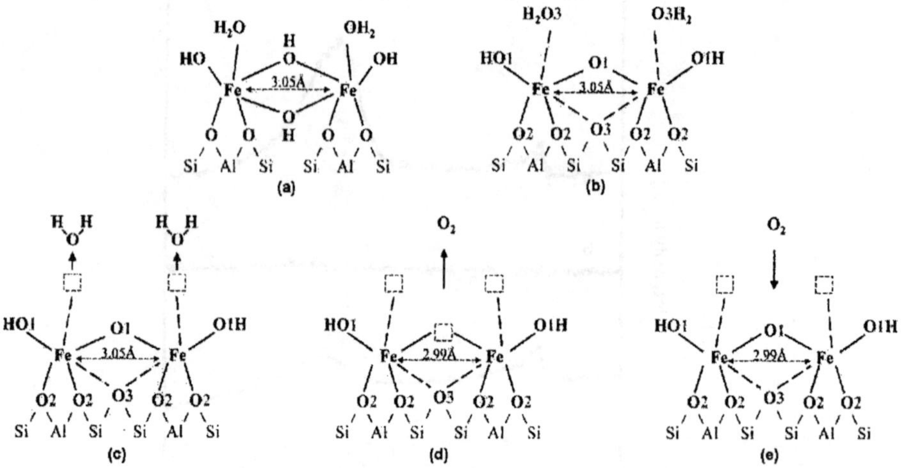

Figure 27 *Proposed structure for the binuclear Fe complexes in Fe-MFI-300 obtained by CVD of $FeCl_3$ upon: (a) washing; (b) mild calcination and treatment in He at 30°C; (c) mild calcination and moderate heating (up to 150°C) in He or O_2/He (50:50): desorption of water; (d) mild calcination and heating in He to 350°C: auto-reduction of iron, ascribed to the removal of the Fe-O-Fe bridging oxygen atom; (e) mild calcination and heating to 350°C in a O_2/He (50:50) gas mixture: the presence of oxygen in the gas phase inhibits the removal of the Fe-O-Fe bridging oxygen atom*[221]

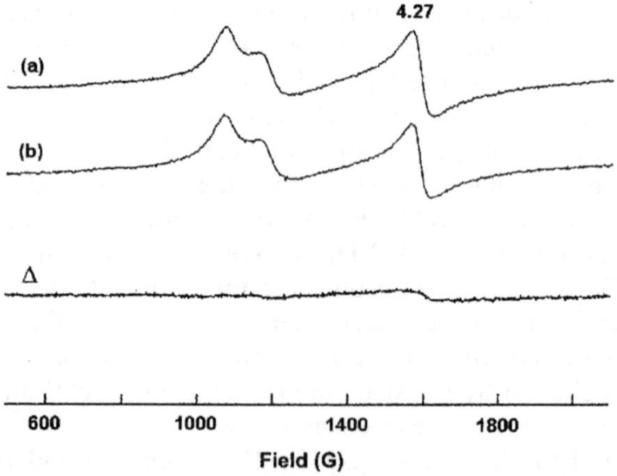

Figure 28 *Low-field part of ESR spectra of Fe-MFI-210 taken at 200°C: (a) dried at 500°C in He + 10% O_2 flow; (b) 20 min after switch to flow of wet He + 10% O_2; (Δ) subtraction of the spectrum (b) from the spectrum (a)*[192]

stable even under high water vapor pressure and without any alteration of the state of the active Fe species. Such substantial studies may be possible to verify their peculiar catalytic characteristics.

Joyner and Stockenhuber[219] have claimed Fe-O nanoclusters possessing an average composition of Fe_4O_4 structures as active species (Figure 29) extensively

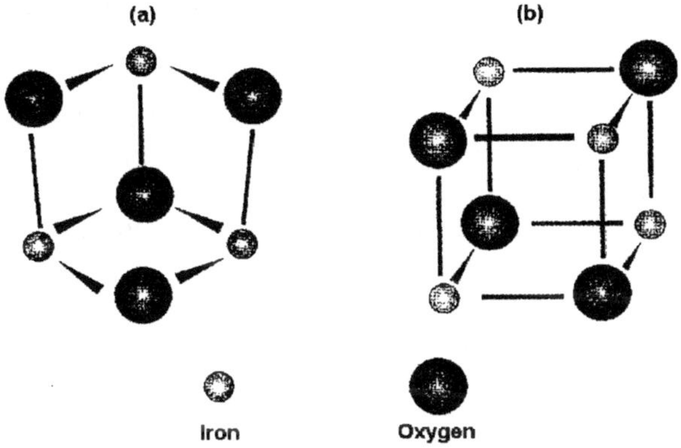

Figure 29 *Possible structural analogues of the iron-oxo nanoclusters formed within the MFI pores: (a) ferredoxin; (b) HIPIP (high-potential iron protein)*[219]

identified by EXAFS, FTIR, XPS and other techniques for Fe-MFI which were obtained from a variety of the preparation methods. Subsequently, isolated iron atoms could be formed, and the extent of the existence of both species on the catalyst surface varied with the preparation routes, particularly with the pretreatment processes. Pretreating the Fe-MFI in an oxygen-free condition leads to the formation of larger amounts of nanoclusters that are quite active for catalytic NO_x reduction under excess oxygen; moreover, the interconversion between divalent Fe and trivalent Fe in these Fe_4O_4 type nanoclusters occurs easily. This indicates that even the CVD technique resulting in catalyst with a variety of Fe sites significantly depends on the activation procedures of the prepared catalyst. Water washing protocols could also alter the individual distribution of Fe species in the zeolite pores.[63] The catalyst undergoing activation treatment in oxygen contained lower amounts of the Fe clusters and revealed lower catalytic performance, suggesting the Fe_4O_4 type nanoclusters may be an active site for HC-SCR reaction. However, it is necessary to investigate a catalyst sample after the reaction in wet stream at an appropriate temperature to understand strong water tolerance and hydrothermal stability of such overexchanged Fe-MFI catalysts. In addition, Lee and Rhee[223] have claimed isolated Fe^{3+} ions as an active reaction site for the present reaction system, and Bell and coworkers[223a] have also proposed that Fe species, in an Fe-MFI catalyst obtained by a solid-state ion exchange technique, exist in the form of isolated cations.

5.4.2. Cu-Exchanged Zeolites. Copper ions and/or complexes exchanged into such commercially manufactured zeolites as MFI, MOR, FAU, FER, BEA and so forth have been shown to be active for deNO_x catalysis with HCs. Catalytic deNO_x activity for this reaction can be maximized when combining copper with the MFI structure zeolites, representatively ZSM-5, depending mainly on the nature of reductant and physicochemical properties of the zeolite employed. These Cu-based zeolites reveal the peak NO_x reduction activity at higher temperatures than supported noble metal catalysts. The challenges in high-temperature $deNO_x$ applications to ICEs are poor hydrothermal durability of the Cu-based zeolites and their sensitivity to the presence of H_2O and SO_2, particularly water vapor (Table 4). The deactivation of the catalysts during the wet $deNO_x$ catalysis under high oxygen pressure has been reported to be reversible and/or irreversible.[11,17,41,48,49,67,111,112,197] The reversibility may be associated predominantly with the competitive adsorption between H_2O and reactant molecules.[46,67,112] On the other hand, the permanent deactivation of the Cu-based catalysts is due to either the destruction of the zeolite support, the adverse geometrical change of the active Cu species in coordination environments, the positions in the zeolite or all of them.[9,21,199] These two distinctive deactivations of the Cu-zeolite catalysts will be discussed below.

To elucidate the reversible activity loss of Cu-MOR catalysts, with relatively low Cu exchange ratio for NO reduction by C_2H_4 and C_3H_6 in dry and wet condition, Kim *et al.*[112] have determined electronic and local structures of synthetic and natural MOR-based Cu catalysts after the reaction at 400°C for 1 h in the presence of 7.3% H_2O. Cu K-edge XANES spectra for both catalysts even after the reaction were basically similar to those of each fresh catalyst. The spectra were quite distinctive compared with the reference samples even for the catalysts exposed to the wet stream, revealing that the copper species are neither Cu_2O, CuO nor $Cu(OH)_2$. A multiple scattering did not appear. It implies that no change in the electronic structure of the Cu ions occurs even after the wet $deNO_x$ catalysis and copper oxide clusters cannot be formed on the surface of the zeolites. Cu K-edge k^3-weighted EXAFS spectra could lead to the same conclusion for the two Cu-exchanged catalysts (Figures 30 and 31). The predominant peak at 1.62Å in cupric oxide was, without a correction for the shift from real distance by a phase shift as stated, assigned to the nearest Cu-O, as compared to 1.50Å in Cu_2O (1.95Å in CuO[224] and 1.85Å in Cu_2O[225]). Either one or two neighboring peaks appear at a longer distance: 2.73Å for Cu_2O, and 2.55 and 3.08Å for CuO. In the case of $Cu(OH)_2$, three prominent peaks are observed at 1.60 and 2.73, and 3.01Å, corresponding to Cu-O and Cu-Cu distances. Both Cu-zeolite catalysts before and after the reaction exhibited only a single peak at 1.54Å, as shown in Figure 31, representing that the copper species are in the form neither of CuO and Cu_2O nor of $Cu(OH)_2$. The 1.54Å peak was assigned to the local structure between the Cu ions and the nearest zeolite framework oxygen, as well-known for Cu-FAU[226,227] and Cu-MFI.[111,197,228,229] These EXAFS spectra clearly indicate no formation of such copper oxides as CuO and Cu_2O, which is quite consistent with the reversible catalyst deactivation. This may be primarily due to the relatively low Cu exchange ratio of the catalyst. It might suggest that the loss of the catalytic activity of the catalyst in the wet condition is caused by another reason.

Kim *et al.*[48,49,112] conducted temperature programmed desorption of H_2O (H_2O-TPD) to elucidate the reversible phenomenon when Cu-MOR catalysts for NO reduction with C_2H_4 and C_3H_6 were periodically exposed to the absence and presence of 7% H_2O. Large amounts of H_2O were adsorbed on the H-MOR and desorbed continuously up to 500°C, as shown in Figure 32, and on Cu-MOR-48

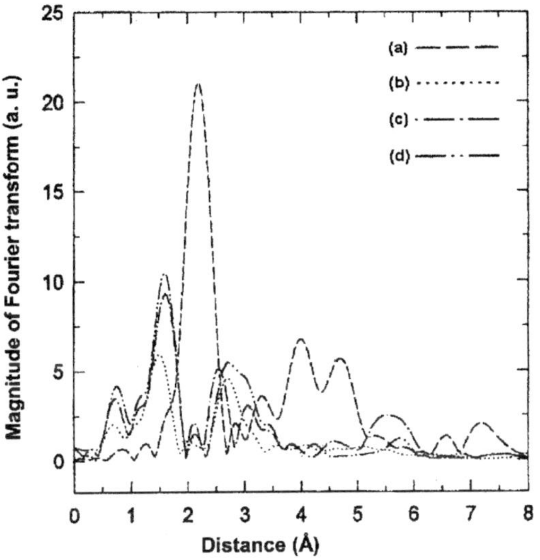

Figure 30 *Cu K-edge EXAFS spectra for (a) Cu foil, (b) Cu_2O, (c) CuO and (d) $Cu(OH)_2$ as standard copper compounds*[112]

up to 380°C. The desorption of H_2O molecules on the natural zeolite-based catalyst appeared at 150, 210, 250 and 460°C, and the total amount of water desorbed was significantly low. It was indicated from the H_2O TPD that the capability of H-MOR to adsorb H_2O and its adsorption strength is much higher than that on Cu-MOR-48 catalyst. The synthetic and natural zeolites possessed the respective Si/Al ratios of 5.2 and 9.2. This distinction may bring the natural zeolite-based Cu catalyst to exhibit strong surface hydrophobicity, which can be supported by earlier results by Flanigen *et al.*[213] who observed that the extent of the water adsorption on an MFI zeolite is inversely related to its Si/Al ratio. The Kubelka-Munk spectra of H_2O adsorbed on the Cu-MORs indicated that water molecules bonded on the Cu ions, depending on the adsorption temperature of H_2O.[112] It indicates that protonic zeolites are not favorable SCR catalyst support for wet deNO_x catalysis, although many HC-SCR reactions have been catalyzed over the H-zeolites.[43,44,47-49,65-69] H_2O ligands coordinated to the active copper species in the zeolite pores may inhibit the adsorption of NO on their surface, thereby leading to a significant decrease of NO chemisorption capacity of the catalyst in the presence of H_2O.

A close correlation between hydrophobicity and competitive adsorption has been verified by the simultaneous adsorption of NO and HC on the catalyst surface with and without 7% H_2O.[49,112] For Cu-MOR-48 catalyst shown in Figure 33, the desorption peaks of NO_x at 110, 160, 210 and 280°C were observed, regardless of the presence of the water vapor; however, the adsorption capacity considerably decreased upon the simultaneous adsorption of NO and H_2O on the catalyst surface. Since the TPD profile of NO adsorbed on H-MOR

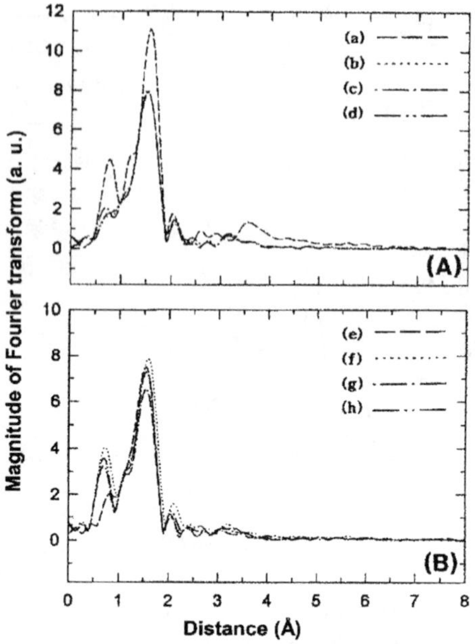

Figure 31 *Cu K-edge EXAFS spectra for (A) Cu-MOR-48 (synthethic MOR) and (B) Cu-MOR-44 (natural MOR) catalysts. (a) and (e) fresh; (b) and (f) after reaction with water at 400°C for 1 h; (c) and (g) after reaction with water at 700°C for 1 h; (d) and (h) after reaction with water at 400°C for 1 h following reaction without water for 1 h. Reaction condition: NO 500 ppm, C_3H_6 2,000 ppm, O_2 4.2 % and 7.3% H_2O*[112]

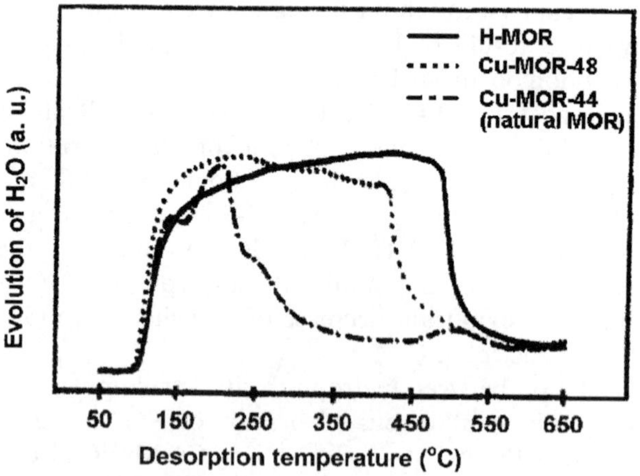

Figure 32 *TPD profiles of H_2O for Cu-MOR catalysts. The ramping rate was 10°C/min, and the carrier (He) flow rate was 40 cm^3/min (adopted from ref. 112)*

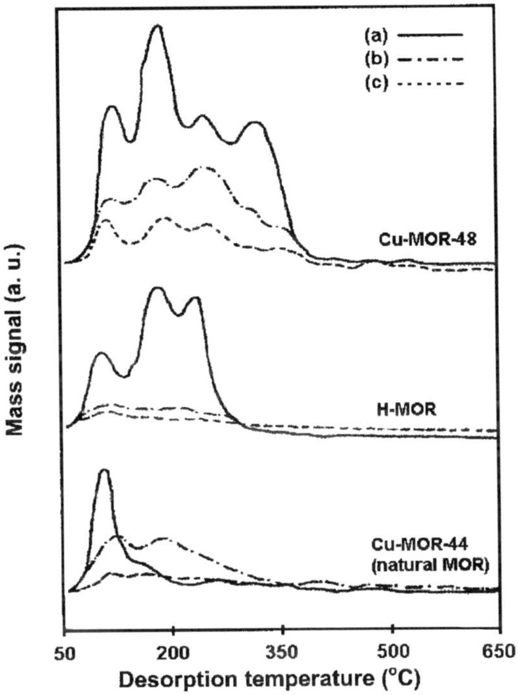

Figure 33 *TPD profiles of NO adsorbed on Cu-MOR catalysts in the absence and presence of H_2O. (a) NO adsorption in the absence of H_2O; (b) summation of nitrogen compounds formed during NO adsorption in the presence of H_2O; (c) NO desorption after simultaneous adsorption of NO and H_2O. The ramping rate was 10°C/min, and the carrier (He) flow rate was 40 cm^3/min (adopted from ref. 112)*

has exhibited peaks at 110, 160 and 210°C, the last one shown for Cu-exchanged catalyst at 280°C is assigned to NO_x species accommodated on the Cu sites which can be strongly inhibited by H_2O during NO adsorption with H_2O. The capability of H-MOR to adsorb NO in a wet stream is significantly low compared to that of Cu-MOR-48 catalyst. The natural zeolite-based Cu-MOR-44 catalyst showed the NO TPD peaks at 100, 146 and 230°C when NO was only adsorbed. Although the simultaneous adsorption of both NO and H_2O on the catalyst led to a severe decrease in the amount of NO adsorbed, the amount of total nitrogen compounds desorbed almost remained the same. This implies that the adsorption sites on the Cu-MOR-44 are less deactivated even in the wet condition. The extent of the decrease of NO adsorption by H_2O well illustrates the water tolerance of the catalyst employed. It suggests that H_2O molecules are competitive with such reactants as NO and HCs to occupy the catalytic reaction sites. These results immediately support the complete reversibility of the deNO_x conversion activity in dry and wet streams as reported for Co-MFI-140, Co-MFI-98, Co-FER-78 and Co-MOR-94,[46,97] Cu-MFI-157,[41] Pd-MOR[162] and Co-FER-98.[51]

Water has affected the adsorption of hydrocarbons C_3H_6 and C_2H_4 on the surface of Cu-zeolite catalysts.[49,67,112] In a representative experiment for C_3H_6 TPD over the natural zeolite-based Cu-NMOR-44 catalyst, the adsorption of C_3H_6 without H_2O revealed two prominent desorption peaks at 100 and 350°C.[67,112] Although the simultaneous adsorption of C_3H_6 and H_2O on the catalyst led to a significant decrease in the amounts of the desorption of C_3H_6 at temperatures ranging from 130 to 250°C, the amounts of C_3H_6 chemisorbed at 350°C maintained even in a wet stream. For the adsorption of C_2H_4 on the catalyst without H_2O, a doublet peak for its desorption below 200°C was observed with a broad one centered at 370°C. However, the intensity of the peaks decreased for the simultaneous adsorption of C_2H_4 and H_2O. By a comparison of the desorption amounts of the both HCs on the catalyst surface from 200 to 500°C, the adsorption capacity of C_3H_6 on the Cu-NMOR-44 was much greater than that of C_2H_4, regardless of the presence of H_2O, and C_3H_6 was predominantly chemisorbed on the catalyst even under the wet condition. This is in good agreement with the observation in which smaller amounts of NO and HCs could be adsorbed on wet surfaces of Cu-MFI-111, Co-MFI-117 and Co-BEA-80[200] as well as of Pd-MFI.[160] Side reactions between HCs and H_2O have also been a minor reason for the deactivation of the MOR-based catalysts for SCR by HCs in a wet feed gas stream.[112] If SO_2 is substantially included in the feed gas stream, their deactivation behavior becomes more complicated.[230,231] These adverse roles of H_2O in lowering the capability of the zeolite catalyst to chemisorb HCs and enhancing such a side reaction should be the main cause for the reversible deactivation of metal-exchanged zeolite catalysts for wet $deNO_x$ reaction. Consequently, the presence of water vapor can inhibit the formation of reaction intermediates due to the competitive adsorption and side reaction occurring during the course of the reaction.[232,233]

The major reason for the irreversible deactivation of Cu-zeolites during HC-SCR reaction in the presence of H_2O vapor could be understood by illustrating the structure and state of the active reaction sites on the catalyst surface. The facile interconversion between Cu^+ and Cu^{2+} in the Cu-zeolites, particularly excessively exchanged ones, depending on such reaction conditions as temperature and the types of reductant, is often considered to be essential for SCR catalysis.[191,234,235] Copper structures bearing extralattice oxygen (ELO) have been proposed to be dimeric $[Cu\text{-}O\text{-}Cu]^{2+}$species[190,191] and small zeolite-hosted copper oxide clusters such as $Cu^{2+}O^-$ or $Cu^{2+}O_2^-$;[236-238] however, the identity of the ELO is still unclear. On the other hand, *in situ* ESR investigations of a Cu-MFI with Cu^{2+}exchange level less than 100% have revealed only the presence of the Cu^{2+},[239] and the SCR activity has been mainly responsible for isolated Cu^{2+} ions on the surface of Cu-ZSM-5 catalysts.[229,240] Recently, Millar *et al.*[241] have assigned three different Cu species on the surface of Cu-MFI with respect to the Cu exchange levels. Underexchanged Cu-MFI catalysts caused the formation of only $Cu^{2+}OH^-(H_2O)$ species bound to two ELO. The Cu species could be transformed into both $Cu^{2+}O^-$ and small CuO clusters during calcination and were assumed to be active catalytic sites for $deNO_x$ reaction. As the exchange level of Cu increased, $Cu^{2+}(H_2O)_2$ complexes coordinated to four ELO were subsequently formed.

Keeping in mind the presence of Cu species mentioned above on the surface of Cu-zeolites, Kharas *et al.*[111] observed irreversible deactivation of Cu-MFI catalyst by H_2O. The performance of Cu-MFI-387 for wet deNO_x catalysis was examined for 1 h at reaction temperatures from 600 to 800°C. The formation of CuO crystallites was detected by EXAFS and XRD measurements for the deactivated Cu catalysts. Substantial deactivation was accompanied by the loss of pore volume during aging, which could also be correlated to the sintering of the Cu entities. The sintering of the copper ions could lead to the destruction of the zeolite structure along with the loss of the catalytic activity during the course of reaction under the LB condition of automotive engine. A similar result has been also observed for Cu-MFI-106 and Cu-MFI-678 catalysts on which Cu ions are transferred to small CuO clusters through the durability test at 500°C for 500 h under a simulated LB exhaust containing 9% H_2O.[197] Dealumination of the catalyst, carbon deposition on the catalyst surface and loss of the catalyst micropore volume hardly occurred by confirming BET surface area, CO chemisorption, NMR and elementary carbon analysis of the catalyst after reaction. Therefore, this indicates that deactivation during the time on-stream operation is due exclusively to the sintering of the active Cu species to very small CuO crystallites that are not active for HC-SCR deNO_x catalysis. This suggests that Cu species in low coordination environments are probably sensitive to water vapor. The solid state ^{27}Al MAS NMR studies by Grinsted *et al.*[143] have revealed a loss of 23% of framework aluminum after wet aging of a Cu-MFI catalyst at 410°C in 10% H_2O in air for 113 h, using ^{27}Al NMR. This can induce agglomeration and redistribution of active Cu complexes to inactive ones, although they have not yet been experimentally verified. The above evidence can still support the permanent activity loss of Cu-zeolite catalysts in the present reaction system. On the other hand, neither the hydrothermal dealumination nor the formation of small CuO particles has been indicated by NMR and XRD measurements for an overexchanged Cu-MFI catalyst treated in a simulated lean mixture at 800°C for 5 h.[242] The deactivation is probably caused by the migration of active Cu^{2+} species to different coordination environments inactive for HC-SCR reaction.

6 Application of HC-SCR DeNO_x Technology to Advanced ICEs

In the previous sections, the water tolerance and the hydrothermal durability of numerous HC-SCR deNO_x catalytic systems and their deactivation behavior have been thoroughly discussed, particularly with respect to the active reaction sites on the catalyst surface for HC- SCR reaction. Prior to proposing potential HC-SCR processes for controlling NO_x emission from advanced engines, one needs to clarify the reason why the HC-SCR technology is still regarded as one of major approaches to reducing NO_x from advanced ICEs in the future. It can be easily expected that the engine performance of advanced ICEs including the emissions of NO_x will be much better in the future than that of the present ones.

It is hoped that the present HC-SCR technology can even meet the future emission regulations. Note that the HC-SCR technology has been recently considered as an inappropriate method mainly due to the rather low NO removal activity in the range of 50% NO conversion.

The demand for advanced diesel and gasoline engines throughout the world is boosted by the need to reduce CO_2 emissions, in addition to the remarkable enhancement of fuel economy. The respective diesel and LB engines can release less CO_2 amounts of *ca.* 20 and 10% than conventional gasoline engines with closed loop TWC systems.[2,12] The exhaust gas from the diesel engines also contains much less amounts of CO and unburned HCs than the stoichiometrically operated conventional engines, although PM emissions from diesel engines are higher, by one to two orders of magnitude, than those from comparable gasoline engines. Suitable exhaust aftertreatment technologies to substantially abate NO_x under LB conditions should be required to utilize such engines with those advantages. Except for HC-SCR technology which is the main topic of the present review, other various approaches, such as the LNC, NSR, so-called NO_x adsorber or NO_x trap, urea-SCR, PAC, and multi-stage catalytic system, have been proposed to reduce NO_x emissions from advanced ICEs, as briefly described in the introductory section. An alternative technology may be a combination of catalytic NO_x decomposition and reduction at the first stage and then catalytic oxidation of residual HC and CO at the final stage;[243] however, it is difficult to remove their emissions during cold-start. Among these technologies investigated for NO_x removal from advanced ICEs, only the NSR technique has been commercialized by Toyota Motors in 1994 for gasoline direct injection (GDI) engine-equipped vehicles to abate NO_x, CO and unburned HCs. Therefore, it is widely anticipated that the NSR approach may be the most viable solution for removing NO_x from LB ICEs to meet future NO_x emission standards such as US Tier 2 and Euro V;[244] however, it has not yet been available for diesel engines.

Major difficulties in applying the NSR systems to diesel deNO_x technology are its weak sulfur tolerance and poor performance of NO_x storage at low temperature.[244,245-247] For on-board regeneration of the NSR catalysts, consisting of Al_2O_3-based Pt and BaO or $BaCO_3$ with other additives, deactivated by SO_2, those systems should periodically undergo high temperatures greater than 650°C[248] but such temperatures cannot be normally achieved even for advanced diesel engines in the future. The exhaust temperatures of many light-duty diesel vehicles fall in the temperature range below 250°C, and NSR catalysts are hardly regenerated at this temperature. Note that the exhaust temperature of many LB gasoline engines is substantially lower, by 150°C, than the stoichiometric ones and similar problems are present. A post-injection technique of diesel fuel to the front position of the catalysts to raise the temperature has been suggested as a way to allow their periodic regeneration;[249] however, this requires a flexible high-pressure fuel injection system for steady-state engine operation without the torque and speed fluctuation.[250] The SO_2 can be easily oxidized to SO_3 over Pt surfaces and the SO_3 reacts with Ba compounds. Unlike nitrates on the NSR catalysts, the sulfates formed are hardly removed from the surface even during a regeneration process. The LB ICEs contain much more O_2 amounts, and the SO_2 oxidation is enhanced over such catalysts. Therefore, a variation of the NSR catalytic systems, *i.e.*, a combination of TWC and NO_x trap catalysts,[251] low oxygen storage TWC and lean NO_x trap[252], is under development to improve the catalyst regeneration and system performance. It should noted that only diesel oxidation catalysts (DOC), and diesel particulate filters (DPF) are currently utilized to reduce PMs, CO and soluble organic fraction (SOF) for recent diesel-driven vehicles.[253]

The lean NO_x catalysts for reducing NO_x from lean-burn gasoline engine-driven vehicles are improper because of their low NO_x conversion, as extensively discussed. Such weak performance is mainly associated with low engine-out HCs concentration after the internal combustion. It suggests that HC-SCR processes could play a proper role in reducing NO_x if subsequent HCs were added to the exhaust stream by using appropriate systems. In 1993, Daimler-Benz in Germany conducted a comparative study for various diesel fuel injection systems to develop future generation passenger-car diesel engines.[254] It was discovered that pilot injection is the key factor in improving noise and meeting future emission standards. Through their joint work with Robert Bosch GmbH since 1994, they developed commercial Common Rail diesel direct injection (CRDDI) engines in 1997. Diesel-fueled commercial vehicles equipped with the CRDDI engines can allow subsequent injection of HCs to the appropriate position between the engine outlet and HC-SCR system, if used, without any fluctuation in the consistency of engine operability. More recently, it has been proposed that the GDI-Homogeneous Charge Compression Ignition (GDI-HCCI) mode offers much lower emissions of NO_x, CO and H_2 but high HC emissions.[255-258] This HCCI combustion technology can provide an engine efficiency as high as compression ignition DDI (CIDDI), which is an advanced version of the commonly-known diesel engines; however, quite lower emissions of PMs comprising nano-particulates are probable through the HCCI combustion. With successful research and development, it can be anticipated that further advanced HCCI-DDI and -GDI engines will be developed in the near future. Such advanced engine technologies including CRDDI, HCCI-CRDDI, GDI, and HCCI-GDI would enable the HC-SCR processes to be one of the most promising approaches for controlling NO_x from advanced ICEs in the future.

7 Summary and Future Direction

Overexchanged Fe-MFI catalysts *via* an oxalate method offer high performance and hydrothermal durability for NO_x reduction with *iso*-butane even in the presence of significant amounts of H_2O and SO_2 for high temperature deNO_x application to the ICEs. However, this approach has hardly been reproducible. On the other hand, the CVD method can prepare such Fe-MFI catalysts on which somewhat lower NO_x removal efficiency but still high durability can be achieved for the present reaction system. Not only could the technique be reproducible for the catalyst preparation, but it also depends strongly on the nature of the parent zeolites and their commercial production history. Unfortunately, the preparation technique requires special care to maximize the formation of the active reaction sites on the catalyst surface; therefore, their facile synthesis method inducing equivalent deNO$_x$ performance and hydrothermal durability as those formed by CVD or anaerobic aqueous exchange of FeC_2O_4, should be primarily developed for future investigation. Overexchanged Fe-zeolite catalysts exhibit peak NO$_x$ removal efficiency around 350°C, and it is frequently recognized that the catalytic system may be suitable for high-temperature deNO$_x$ technology. However, the peak temperature is too low to employ for such deNO$_x$ applications to current LB gasoline and advanced GDI engines. Consequently, it is suggested that under floor catalyst (UFC) systems are preferable for the Fe-exchanged zeolite catalysts rather than close-coupled catalyst (CCC) ones being currently used. Other future interest in deNO$_x$ catalysis over overexchanged Fe-zeolite catalysts may be to prevent the formation of byproducts including HCN, which has been produced up to *ca.* 150 ppm,[136,259] and to enhance the ability of the catalysts to completely oxidize CO, which is always produced in significant amounts during the course of the reaction.

Co-exchanged zeolites are sensitive to the presence of water vapor, depending strongly on the preparation route and the characteristics of the parent zeolite. Co-zeolites including Co-BEA, Co-based bimetallic and overexchanged Co-MFI prepared *via* CVD route, may offer strong water tolerance and hydrothermal stability for deNO$_x$ reaction with realistic amounts of H_2O in the feed gas stream. However, such a high performance of the catalysts for the wet deNO$_x$ catalysis could be obtained at high temperatures (at least $>$ 400°C). Most of the supported metal oxides exhibit poor water tolerance and their hydrothermal stability is not well documented. The deNO$_x$ activity of supported Ag catalyst strongly depends on the catalyst supports employed. Al_2O_3-supported Ag catalyst offers the best performance for NO removal reaction and possesses a wider temperature window than supported Pt catalysts. Although the catalyst is strongly inhibited by the presence of H_2O at temperatures lower than 400°C, it still has a potential as a high-temperature deNO$_x$ catalyst for LB gasoline engines as for the Co-based catalysts. The Ag catalyst can be prepared by using relatively simpler techniques including incipient wetness and so-gel synthesis. Contrary to the latter method, the incipient wetness technique has difficulty in preparing high Ag-loaded catalyst with excellent dispersion desirable for higher NO removal activity. The Ag catalyst reveals better activity maintenance during wet deNO$_x$ catalysis when oxygenated HCs, representatively alcohols, are employed as a reductant. However, on-board use of such reductants, one of the major challenges for future study, should be resolved for its commercialization. Furthermore, the catalyst was severely deactivated by SO_2; for instance, about 70% of the initial activity for an Ag/Al_2O_3 catalyst decreased to 15% after 50 h of time on-stream operation in the subsequent presence of 50 ppm SO_2.[260] Therefore, an approach to improve the sulfur tolerance of such Ag catalysts should be another future challenge.

Supported Pd catalysts exhibit unstable time on-stream activity even for dry deNO$_x$ reaction. Supported Pt catalysts appear to maintain sufficient water tolerance and hydrothermal durability under realistic conditions containing H_2O and SO_2. They are particularly active at reaction temperatures below 250°C; therefore, these SCR catalysts can be utilized for low-temperature SCR technology for reducing NO_x from advanced CRDDI engines among LB ICEs. Such engine-equipped vehicles readily allow the use of on-board HCs as a subsequent reductant. However, these catalysts still contain future challenges to be resolved: high selectivity toward N_2O production by $S_{N_2O} = 40 - 60\%$, narrow operating window ($\leqslant$ 30°C) and relatively easier oxidation of SO_2 to SO_3. Most of the HC-SCR reactions over the supported Pt catalysts produce large amounts of N_2O, which is known to be one of the greenhouse gases, besides the catalytic process by toluene as a reductant. For a future attempt to overcome this problem, an alternative approach may be the reaction system containing NO-CO-H_2 over supported Pt catalysts in the presence of excess O_2 and H_2O. In general, vehicle exhaust contains both H_2 and CO that may also participate to remove NO. For high deNO_x performance of the catalysts to meet future diesel NO_x emission standards, diesel fuel can be subsequently added through the post-injection method as already discussed for modern CRDDI engines. However, the direct use of diesel fuel as a reductant may cause less improvement of the catalytic performance, and an excessive amount of fuel injection may be required. If diesel fuel were converted selectively to CO and H_2, which are generally known as effective molecules for NO removal reaction through reforming processes, the injection of diesel to SCR reactor offers twofold advantages. Recently, the selectivity of NO to N_2 has reached to about 90% at 250°C for the NO-CO-H_2 reaction system over a 0.5 wt.% Pt/Al_2O_3 catalyst.[261]

Cu-exchanged zeolite catalysts also possess high deNO_x activity and are the most extensively investigated catalyst so far; however, these HC-SCR catalysts still exhibit poor water tolerance and hydrothermal durability, as confirmed by a huge body of studies, which are essential for deNO_x applications to such ICEs as LB gasoline- and diesel-equipped vehicles. The catalyst reveals either temporary or permanent deactivation for SCR reaction in the presence of water vapor. The reversible deactivation is mainly associated with the competitive adsorption of both H_2O and reactants on the active reaction sites, Cu species, while the irreversible activity loss is due either to destruction of the zeolite support, adverse geometrical change of the active Cu species in coordination environments and positions in the structure of zeolite, or all of them. Generally, the use of Cu-zeolites containing low Si/Al ratios offers better water tolerance and hydrothermal durability at the expense of catalytic activity. Developing a better Cu-based catalyst is still an open challenge as a future effort for NO_x emission control from advanced ICEs, particularly efficient lean-burn gasoline engines which promise a 10% gain in fuel efficiency. A chemical vapor deposition technique still allows high NO_x conversion activity and excellent hydrothermal excursion of overexchanged Fe-MFI catalysts. Moreover, better stability during deNO_x catalysis under high-temperature steam could be obtained for a Co-MFI catalyst prepared *via* the CVD route. Therefore, it is of particular future interest to prepare Cu-zeolites *via* the CVD technique for wet deNO_x catalysis at temperatures greater than at least 250°C. This approach may be a way to prepare

Cu-exchanged zeolite catalyst containing strong water tolerance and hydrothermal stability, although a method for preparation by using a solid ion exchange chemical transport technique[262,263] has been suggested.

Acknowledgment

One of the authors (MHK) would like to thank Mr. Y.-I. Song in the Department of Emissions Research at Advanced Technology Center of Hyundai Motor Company for providing useful references and helpful discussion on the recent development of automotive engine aftertreament technology.

Appendix

A. Zeolite structure code

BEA = zeolite β
FAU = faujasite
FER = ferrierite
HEU = heulandite, clinoptilolite
LTL = zeolite L
MFI = ZSM-5
MFO = MCM-41
MOR = mordenite

B. Acronym

AFR: air/fuel ratio
CAES: corporate average emission standard
CIDDI: compression ignition diesel direct injection
CRDDI: Common Rail diesel direct injection
CVD: chemical vapor deposition
DOC: diesel oxidation catalyst
DPF: diesel particulate filter
DRIFTS: diffuse reflectance infrared Fourier transform spectroscopy
ELO: extralattice oxygen
EPR: electron paramagnetic resonance
ESR: electron spin resonance
EXAFS: extended X-ray absorption fine structure
GDI: gasoline direct injection
GVWR: gross vehicle weight rating
HCCI: homogeneous charge compression ignition
HDDV: heavy-duty diesel vehicle
HDV: heavy-duty vehicle
HLDT: heavy light-duty truck
HRAEM: high resolution analytical electron microscopy

ICE: internal combustion engine
LB: lean burn
LDV: light-duty vehicle
LDT: light-duty truck
LLDT: light light-duty truck
LNC: lean NO_x catalyst
MAS NMR: magic angle spinning nuclear magnetic resonance
MDPV: medium-duty passenger vehicle
NSR: NO_x storage and reduction
PAC: plasma-assisted catalysis
SCR: selective catalytic reduction
SOF: soluble organic fraction
SET: supplemental emission test
SFTP: supplemental federal test procedure
TEM: transmission electron microscopy
TPD: temperature programmed desorption
TPR: temperature programmed reduction
TWC: three-way catalytic converter
XANES: X-ray absorption near edge structure
XAS: X-ray absorption spectroscopy
XPS: X-ray photoelectron spectroscopy

References

1. A. Fritz and V. Pitchon, *Appl. Catal. B*, 1997, **13**, 1.
2. K. C. Taylor, Catalysis-Science and Technology (J. R. Anderson and M. Boudart, Eds., Vol. 5), Springer-Verlag, Berlin, 1984, p.119.
3. T. V. Johnson, *SAE* 010184, 2001.
4. A. Amberntsson, M. Skoglundh, S. Ljungstrom and E. Fridell, *J. Catal.*, 2003, **217**, 253.
5. G. Fornasari, F. Trifiro, A. Vaccari, F. Prinetto, G. Ghiotti and G. Centi, *Catal. Today*, 2002, **75**, 421.
6. P. G. Blakeman, G. R. Chandler, G. A. John, A. J. J. Wilkins, *SAE* 013624, 2001.
7. J. Hoard, *SAE* 010185, 2001.
8. S. J. Schmieg, B. K. Cho and S. H. Oh, *SAE* 013565, 2001.
9. M. Shelef, *Chem. Rev.*, 1995, **95**, 209.
10. T. Kreuzer, E. S. Lox, D. Lindner and J. Leyrer, *Catal. Today*, 1996, **29**, 17 and therein.
11. S. Y. Chung, S.-H. Oh, M. H. Kim, I.-S. Nam and Y. G. Kim, *Catal. Today*, 1999, **54**, 521.
12. J. Leyrer, E. S. Lox and W. Strehlau, *SAE* 952495, 1995.
13. W. O. Siegl, R. H. Hammerle, H. M. Herrmann, B. W. Wenclawiak and B. Luers-Jongen, *Atmospheric Environ.*, 1999, **33**, 797.
14. W. Held, A. Konig, T. Richter and L. Puppe, *SAE* 900496, 1990.
15. M. Iwamoto, *Proceedings of Meeting on Catalytic Technology for Removal of Nitrogen Monoxide*, Tokyo, Japan, January, 1990, p. 17.
16. T. J. Truex, R. A. Searles and D. C. Sun, *Platinum Metals Rev.*, 1992, **36**, 2.

17. M. Iwamoto and N. Mizuno, *J. Auto. Eng.*, 1993, **207**, 23.
18. R. H. H. Smits and Y. Iwasawa, *Appl. Catal. B*, 1995, **6**, L201.
19. B. J. Adelman, T. Beutel, G.-D. Lei and W. M. H. Sachtler, *Appl. Catal. B*, 1996, **11**, L1.
20. T. Tabata, M. Kokitsu and O. Okada, *Catal. Today*, 1994, **22**, 147.
21. V. I. Parvulescu, P. Grange and B. Delmon, *Catal. Today*, 1998, **46**, 233.
21a. B. I. Bertelsen, *Top. Catal.*, 2001, **16/17**, 15.
22. G. B. Weller, 2007 Technology Primer, Clean Diesel Engine Implementation Workshop, Chicago, USA, August 6 – 7, 2003.
22a. J. G. E. Cohn, W. Orange, D. R. Steele and H. C. Anderson, *US Patent* 2975025, 1961.
23. O. J. Adlhart, S. G. Hindin and R. E. Kenson, *Chem. Eng. Prog.*, 1971, **67**, 73.
24. R. J. Ayen and M. S. Peters, *Ind. Eng. Chem. Proc. Des. Dev.*, 1962, **1**, 205.
25. A. Sotoodehnia-Korrani and K. Nobe, *Ind. Eng. Chem. Proc. Des. Dev.*, 1970, **9**, 326.
26. M. Shelef and J. T. Kummer, *Chem. Eng. Progr. Symposium Ser.*, 1971, **67**, 74.
27. J. W. Ault and R. J. Ayen, *AIChEJ*, 1971, **17**, 265.
28. J. C. Vartuli and R. D. Gonzalez, *J. Catal..*, 1974, **32**, 470.
29. Y. Murakami, K. Hayashi, K. Yasuda, T. Ito, T. Minami and A. Miyamoto, *Nippon Kagaku Kaishi*, 1977, 173.
30. Y.-H. Hu and J. W. Hightower, *Symposium on Emission Control Catalysts*, San Francisco, USA, August 29-September 3, 1976: *Preprints, ACS Div. Petro. Chem.* 1976, **21**, 841.
31. J. R. Hardee and J. W. Hightower, *J. Catal.*, 1984, **86**, 137.
32. W. Held and A. Konig, *Deutsch Patent* 3642018 A1, 1987.
33. W. Held, A. Konig and L. Puppe, *Japanese Patent* 63283727 A, 1988.
34. W. Held, A. Konig and L. Puppe, *Deutsch Patent* 3713169 A1, 1988.
35. W. Held, A. Konig and L. Puppe, *European Patent* 0286967 A2, 1988.
36. Y. Fujitani, H. Muraki, S. Kondo and M. Fukui, *Japanese Patent* 63100919 A, 1988.
37. Y. Fujitani, H. Muraki, S. Kondo and M. Fukui, *Deutsch Patent* 3735151 A1, 1988.
38. T. Tanaka, S. Matsumoto, H. Muraki and S. Kondo, *Japanese Patent* 1130735 A, 1989.
39. M. Iwamoto, H. Yahiro, K. Tanda, N. Mizuno, Y. Mine and S. Kagawa, *J. Phys. Chem.*, 1991, **95**, 3727.
40. M. Iwamoto and H. Hamada, *Catal. Today*, 1991, **10**, 57.
41. M. Iwamoto, N. Mizuno and H. Yahiro, *Proceedings of 10th International Congress on Catalysis*, Budapest, Hungry, July 19 – 24, 1992, p. 213.
42. M. Iwamoto, H. Yahiro, S. Shundo, Y. Yu-u and N. Mizuno, *Appl. Catal.*, 1991, **69**, L15.
43. S. Sato, Y. Yu-u, H. Yahiro, N. Mizuno and M. Iwamoto, *Appl. Catal.*, 1991, **70**, L1.
44. S. Sato, H. Hirabayashi, H. Yahiro, N. Mizuno and M. Iwamoto, *Catal. Lett.*, 1992, **12**, 193.
45. Y. Teraoka, H. Ogawa, H. Furukawa and S. Kagawa, *Catal. Lett.*, 1992, **12**, 361.
46. Y. Li, P. J. Battavio and J. N. Armor, *J. Catal.*, 1993, **142**, 561.
47. M. H. Kim, I.-S. Nam and Y. G. Kim, *Hwahak Konghak* (*in Korean*), 1994, **32**, 402.
48. M. H. Kim, I.-S. Nam and Y. G. Kim, *Appl. Catal. B*, 1995, **6**, 297.
49. M. H. Kim, I.-S. Nam and Y. G. Kim, *Stud. Surf. Sci. Catal.*, 1997, **105**, 1493.
50. H. Ohtsuka and T. Tabata, *Appl. Catal. B*, 1999, **21**, 133.
51. D. I. Lee, T. J. Lee, S.-W. Ham, I.-S. Nam, Y. S. Oh and Y.-S. Baek, *Theor. Appl. Chem. Eng.* (*in Korean*), 2001, **7**, 225.

52. X. Feng and W. K. Hall, *Catal. Lett.*, 1996, **41**, 45.
53. X. Feng and W. K. Hall, *J. Catal.*, 1997, **166**, 368.
54. H.-Y. Chen and W. M. H. Sachtler, *Catal. Lett.*, 1998, **50**, 125.
55. H.-Y. Chen and W. M. H. Sachtler, *Catal. Today*, 1998, **42**, 73.
56. T. V. Voskoboinikov, H. -Y. Chen and W. M. H. Sachtler, *Appl. Catal. B*, 1998, **19**, 279.
57. H.-Y. Chen, T. V. Voskoboinikov and W. M. H. Sachtler, *J. Catal.*, 1998, **180**, 171.
58. H.-Y. Chen, T. V. Voskoboinikov and W. M. H. Sachtler, *J. Catal.*, 1999, **186**, 91.
59. H.-Y. Chen, T. V. Voskoboinikov and W. M. H. Sachtler, *Catal. Today*, 1999, **54**, 483.
60. P. Marturano, A. Kogelbauer and P. Prins, *J. Catal.*, 2000, **190**, 460.
61. S. I. Woo, D. K. Kim, Y. K. Park, M. R. Kim and P. Decyk, *Catal. Lett.*, 2003, **69**, 85.
62. F. Heinrich, C. Schmidt, E. Loffler and W. Grunert, *Catal. Commun.*, 2001, **2**, 317.
63. F. Heinrich, C. Schmidt, E. Loffler, M. Menzel and W. Grunert, *J. Catal.*, 2002, **212**, 157.
64. X. Wang, H.-Y. Chen and W. M. H. Sachtler, *J. Catal.*, 2001, **197**, 281.
65. H. Hamada, Y. Kintaichi, M. Sasaki, T. Ito and M. Tabata, *Appl. Catal.*, 1990, **64**, L1.
66. H. Hamada, Y. Kintaichi, M. Sasaki, T. Ito and M. Tabata, *Appl. Catal.*, 1991, **75**, L1.
67. M. H. Kim, *PhD Dissertation, Effect of Water Vapor and SO_2 on the Selective Reduction of NO_x over Mordenite-Type Zeolite Catalysts by Hydrocarbons*, Pohang University of Science and Technology, 1996.
68. T. Nanba, A. Obuchi, H. Izumi, Y. Sugiura, J. Xu, J. Uchisawa and S. Kushiyama, *J. Chem. Soc., Chem. Commun.*, 173, 2001.
69. E. E. Miro, G. Imoberdorf, J. Vassallo and J. O. Petunchi, *Appl. Catal. B*, 1999, **22**, 305.
70. H. Hamada, *Catal. Today*, 1994, **22**, 21.
71. A. Bahamonde, C. Knapp, P. Avila and J. Blanco, *Appl. Catal. B*, 1998, **19**, 1.
72. M. D. Amiridis, K. L. Roberts and C. J. Pereira, *Appl. Catal. B*, 1997, **14**, 203.
73. J. Perez-Ramirez, J. M. Garcia-Cortes, F. Kapteijn, G. Mul, J. A. Moulijn, C. Salinas-Martinez de Lecea, *Appl. Catal. B*, 2001, **29**, 285.
74. S. Lacombe, J. H. B. J. Hoebink and G. B. Marin, *Appl. Catal. B*, 1997, **12**, 207.
75. R. Burch and D. Ottery, *Appl. Catal. B*, 1997, **13**, 105.
76. E. Seker and E. Gulari, *J. Catal.*, 2000, **194**, 4.
77. H. Ohtsuka, T. Tabata and T. Hirano, *Appl. Catal. B*, 2000, **28**, L73.
78. M. Konsolakis and I. V. Yentekakis, *J. Catal.*, 2001, **198**, 142.
79. A. Gervasini, P. Carniti and V. Ragaini, *Appl. Catal. B*, 1999, **22**, 201.
80. A. A. Nikolopoulos, E. S. Stergioula, E. A. Efthimiadis and I. A. Vasalos, *Catal. Today*, 1999, **54**, 439.
81. M. Iwamoto, A. M. Hernandez and T. Zengyo, *J. Chem. Soc., Chem. Commun.*, 1997, 37.
82. R. Burch and A. Ramli, *Appl. Catal. B*, 1998, **15**, 49.
83. H. Hosose, H. Yahiro, N. Mizuno and M. Iwamoto, *Chem. Lett.*, 1991, 1859.
84. Y. Torikai, H. Yahiro, N. Mizuno and M. Iwamoto, *Catal. Lett.*, 1991, **9**, 91.
85. T. Miyadera, *Appl. Catal. B*, 1993, **2**, 199.
86. T. Miyadera, *Appl. Catal. B*, 1997, **13**, 157.
87. T. Miyadera, *Appl. Catal. B*, 1998, **16**, 155.
88. F. C. Meunier and J. R. H. Ross, *Appl. Catal. B*, 2000, **24**, 23.

89. T. Furusawa, L. Lefferts, K. Seshan and K. Aika, *Appl. Catal. B*, 2003, **42**, 25.
90. Y. Kintaichi, H. Hamada, M. Tabata, M. Sasaki and T. Ito, *Catal. Lett.*, 1990, **6**, 239.
91. T. Maunula, J. Ahola and H. Hamada, *Appl. Catal. B*, 2000, **26**, 173.
92. T. Ohno, Y. Bunno, F. Hatayama, Y. Toda and H. Miyata, *Appl. Catal. B*, 2001, **30**, 421.
93. K. K. Hansen, E. M. Skou, H. Christensen and T. Turek, *J. Catal.*, 2001, **199**, 132.
94. A. Subbiah, B. K. Cho, R. J. Blint, A. Gujar, G. L. Price and J. E. Yie, *Appl. Catal. B*, 2003, **42**, 155.
95. Y. Li and J. N. Armor, *Appl. Catal. B*, 1992, **1**, L31.
96. Y. Li and J. N. Armor, *J. Catal.*, 1994, **150**, 376.
97. Y. Li and J. N. Armor, *Appl. Catal. B*, 1993, **3**, L1.
98. F. Witzel, G. A. Sill, G and W. K. Hall, *J. Catal.*, 1994, **149**, 229.
99. T. J. Lee, I.-S. Nam, S.-W. Ham, Y.-S. Baek and K.-H. Shin, *Appl. Catal. B*, 2003, **41**, 115.
100. A. Shichi, A. Satsuma, M. Iwase, K.-I. Shimizu, S.-I. Komai and T. Hattori, *Appl. Catal. B*, 1998, **17**, 107.
101. K. Yogo, M. Ihara, I. Terasaki and E. Kikuchi, *Catal. Lett.*, 1993, **17**, 303.
102. T. Tabata, M. Kokitsu and O. Okada, *Catal. Lett.*, 1994, **25**, 393.
103. K. Yogo, M. Umeno, H. Watanabe and E. Kikuchi, *Catal. Lett.*, 1993, **19**, 131.
104. M. C. Demicheli, L. C. Hoang, J. C. Menezo, J. Barbier and M. Pinabiau-Carlier, *Appl. Catal. A*, 1993, **97**, L11.
105. X. Zhang, A. B. Walters and M. A. Vannice, M. A., *J. Catal.*, 1994, **146**, 568.
106. H. Ohtsuka and T. Tabata, *Appl. Catal. B*, 2000, **26**, 275.
107. J. M. Ramallo-Lopez, F. G. Requejo, L. B. Gutierrez and E. E. Miro, *Appl. Catal. B*, 2001, **29**, 35.
108. J. A. Sullivan and J. Cunningham, *Appl. Catal. B*, 1998, **15**, 275.
109. D. Pietrogiacomi, S. Tuti, M. C. Campa and V. Indovina, *Appl. Catal. B*, 2000, **28**, 43.
110. X. Wang, H.-Y. Chen and W. M. H. Sachtler, *Appl. Catal. B*, 2001, **29**, 47.
111. K. C. C. Kharas, H. J. Robota and D. J. Liu, *Appl. Catal. B*, 1993, **2** , 225.
112. M. H. Kim, I.-S. Nam and Y. G. Kim, *Appl. Catal. B*, 1997, **12**, 125.
113. J. A. Z. Pieterse, R. W. van den Brink, S. Booneveld and F. A. de Bruijn, *Appl. Catal. B*, 2002, **39**, 167.
114. P. Thormahlen, E. Fridell, N. Cruise, M. Skoglundh and A. Palmqvist, *Appl. Catal. B*, 2001, **31**, 1.
115. L. Chen, T. Horiuchi, T. Osaki and T. Mori, *Appl. Catal. B*, 1999, **23**, 259.
116. K. Sato, T. Fujimoto, S. Kanai, Y. Kintaichi, M. Inaba, M. Haneda and H. Hamada, *Appl. Catal. B*, 1997, **13**, 27.
117. J. R. Gonzalez-Velasco, J. A. Botas, J. A. Gonzalez-Marcos and M. A. Gutierrez-Ortiz, *Appl. Catal. B*, 1997, **12**, 61.
118. C. Descorme, P. Gelin, C. Lecuyer and M. Primet, *J. Catal.*, 1998, **177**, 352.
119. M. Ogura, S. Kage, T. Shimojo, J. Oba, M. Hayashi, M. Matsukata and E. Kikuchi, *J. Catal.*, 2002, **211**, 75.
120. G. Mabilon and D. Durand, *Catal. Today*, 1993, **17**, 285.
121. R. Gopalakrishnan, P. R. Stafford, J. E. Davidson, W. C. Hecker and C. H. Bartholomew, *Appl. Catal. B*, 1993, **2**, 165.
122. R. Burch and S. Scire, *Appl. Catal. B*, 1994, **3**, 295.
123. K. A. Bethke and M. C. Kung, *J. Catal.*, 1997, **172**, 93.
124. W. Li, M. Sirilumpen and R. T. Yang, *Appl. Catal. B*, 1997, **11**, 347.

125. A. Ueda, T. Oshima and M. Haruta, *Appl. Catal. B*, 1997, **12**, 81.
126. M. Misono, Y. Hirao and C. Yokoyama, *Catal. Today*, 1997, **38**, 157.
127. T. Maunula, Y. Kintaichi, M. Inaba, M. Haneda, K. Sato and H. Hamada, *Appl. Catal. B*, 1998, **15**, 291.
128. K.-I. Shimizu, A. Satsuma and T. Hattori, *Appl. Catal. B*, 1998, **16**, 319.
129. K.-I. Shimizu, H. Maeshima, A. Satsuma and T. Hattori, *Appl. Catal. B*, 1998, **18**, 163.
130. X. Wang, T. Zhang, X. Sun, W. Guan, D. Liang and L. Lin, *Appl. Catal. B*, 2000, **24**, 169.
131. M. Haneda, Y. Kintaichi, T. Mizushima, N. Kakuta and H. Hamada, *Appl. Catal. B*, 2001, **31**, 81.
132. E. Kikuchi, K. Yogo, S. Tanka and M. Abe, *Chem. Lett.*, 1991, 1063.
133. S.-C. Shen and S. Kawi, *Appl. Catal. B*, 2003, **45**, 63.
134. T. Horiuchi, T. Fujiwara, L. Chen, K. Suzuki and T. Mori, *Catal. Lett.*, 2002, **78**, 319.
135. I. D. Lick, A. Carrascull, M. Ponzi and E. N. Ponzi, *Catal. Lett.*, 2003, **89**, 179.
136. I. O. Y. Liu, N. W. Cant, M. Kogel and T. Turek, *Catal. Lett.*, 1999, **63**, 241.
137. M. Serban, I. Halasz and R. Datta, *Catal. Lett.*, 1999, **63**, 217.
138. E. A. Efthimiadis, G. D. Lionta, S. C. Christoforou and I. A. Vasalos, *Catal. Today*, 1998, **40**, 15.
139. A. A. Battiston, J. H. Bitter and D. C. Koningsberger, *J. Catal.*, 2003, **218**, 163.
140. M. Kogel, R. Monning, W. Schwieger, A. Tissler and T. Turek, *J. Catal.*, 1999, **182**, 470.
141. W. K. Hall, X. Feng, J. Dumesic and R. Watwe, *Catal. Lett.*, 1998, **52**, 13.
142. M. Konno, T. Chikahisa, T. Murayama and M. Iwamoto, *SAE* 920091, 1992.
143. R. A. Grinsted, H. W. Zhen, C. Montreuil, M. J. Rokosz and M. Shelef, *Zeolite*, 1993,**13**, 602.
144. T. Tabata, H. Ohtsuka, O. Okada, L. M. F. Sabatino and G. Bellussi, *Micropor. Mesopor. Mat.*, 1998, **21**, 517.
145. Y. Li and J. N. Armor, *Appl. Catal. B*, 1993, **2**, 239.
146. M. C. Campa, S. De. Rossi, G. Ferraris and V. Indovina, *Appl. Catal. B*, 1996, **8**, 315.
147. J. A. Z. Pieterse, R. W. van den Brink, S. Booneveld and F. A. de Bruijn, *Appl. Catal. B*, 2003, **46**, 239.
148. S. E. Maisuls, K. Seshan, S. Feast and J. A. Lercher, *Appl. Catal. B*, 2001, **29**, 69.
149. L. B. Gutierrez, A. V. Boix, E. A. Lombardo and J. L. G. Fierro, *J. Catal.*, 2001, **199**, 60.
150. D. Kaucky, A. Vondrova, J. Dedecek and B. Wichterlova, *J. Catal.*, 2000, **194**, 318.
151. J. R. Regalbuto, T. Zheng and J. T. Miller, *Catal. Today*, 1999, **54**, 495.
152. K. Kagawa Y. Ichikawa, S. Iwamoto and T. Inui, *Catal. Lett.*, 1998, **52**, 145.
153. M. C. Campa, I. Luisetto, D. Pietrogiacomi and V. Indovina, *Appl. Catal. B*, 2003, **46**, 511.
154. X. Wang, H.-Y. Chen and W. M. H. Sachtler, *Appl. Catal. B*, 2000, **26**, L227.
155. A. Jentys, A, Lugstein and H. Vinek, *J. Chem. Soc., Faraday Trans.*, 1997, **93**, 4091.
156. R. S. da Cruz, A. J. S. Mascarenhas and H. M. C. Andrade, *Appl. Catal. B*, 1998, **18**, 223.
157. S.-J. Jong and S. Cheng, *Appl. Catal. A*, 1995, **126**, 51.
158. H. Ohtsuka, T. Tabata, O. Okada, L. M. F. Sabatino and G. Bellussi, *Catal. Lett.*, 1997, **44**, 265.
159. M. Ogura, S. Kage, M. Hayashi, M. Matsukata and E. Kikuchi, *Appl. Catal. B*, 2000, **27**, L213.

160. M. Ogura, M. Hayashi, S. Kage, M. Matsukata and E. Kikuchi, *Appl. Catal. B*, 1999, **23**, 247.
161. C. Descorme, P. Gelin, C. Lecuyer and M. Primet, *Appl. Catal. B*, 1997, **13**, 185.
162. H. Uchida, K.-I. Yamaseki and I. Takahashi, *Catal. Today*, 1996, **29**, 99.
163. B. J. Adelman and W. M. H. Sachtler, *Appl. Catal. B*, 1997, **14**, 1.
164. Z. Sobalik, J. Dedecek, L. Ikonnikov and B. Wichterlova, *Micropor. Mesopor. Mat.*, 1998, **21**, 525.
165. L. Drozdova, R. Prins, J. Dedecek, Z. Sobalik and B. Wichterlova, *J. Phys. Chem. B*, 2002, **106**, 2240.
166. Y. Li and J. N. Armor, *J. Catal.*, 1994, **145**, 1.
167. C. Rottlander, R. Andorf, C. Plog, B. Krutzsch and M. Baerns, *Appl. Catal. B*, 1996, **11**, 49.
168. J. Guo, M. Konno, T. Chikahisa, T. Murayama and M. Iwamoto, *JSAE Rev.*, 1995, **16**, 21.
169. H. K. Shin, H. Hirabayashi, H. Yahiro, M. Watanabe and M. Iwamoto, *Catal. Today*, 1995, **26**, 13.
170. T. Sowade, C. Schmidt, F.-W. Schutze, H. Berndt and W. Grunert, *J. Catal.*, 2003, **214**, 100.
171. Y. Traa, B. Burger and J. Weitkamp, *Micropor. Mesopor. Mat.*, 1999, **30**, 3.
172. A. Obuchi, A. Ohi, M. Nakamura, A. Ogata, K. Mizuno and H. Obuchi, *Appl. Catal. B*, 1993, **2**, 71.
173. R. Burch, P. J. Millington and A. P. Walker, *Appl. Catal. B*, 1994, **4**, 65.
174. J. M. Garcia-Cortes, M. J. Illan-Gomez, A. L. Solano and C. Salinas-Martinez de Lecea, *Appl. Catal. B*, 2000, **25**, 39.
175. E. Seker and E. Gulari, *J. Catal.*, 1998, **179**, 339.
175a. R. Burch and D. Ottery, *Appl. Catal. B*, 1996, **9**, L19.
176. R. Burch and P. J. Millington, *Catal. Today*, 1995, **26**, 185.
177. R. Burch, *Catal. Today*, 1997, **35**, 27.
178. P. Denton, A. Giroir-Fendler, H. Praliaud and M. Primet, *J. Catal.*, 2000, **189**, 410.
179. G. R. Bamwenda, A. Ogata, A. Obuchi, J. Oi, K. Mizuno and J. Skrzypek, *Appl. Catal. B*, 1995, **6**, 311.
180. G. Zhang, T. Yamaguchi, H. Kawakami and T. Suzuki, *Appl. Catal. B*, 1992, **1**, L15.
181. J.-H. Lee and H. H. Kung, *Catal. Lett.*, 1998, **51**, 1.
182. A. Obuchi, I. Kaneko, J. Oi, A. Ohi, A. Ogata, G. R. Bamwenda and S. Kushiyama, *Appl. Catal. B*, 1998, **15**, 37.
183. A. Obuchi, I. Kaneko, J. Uchisawa, A. Ohi, A. Ogata, G. R. Bamwenda and S. Kushiyama, *Appl. Catal. B*, 1998, **19**, 127.
184. V. Pitchon and A. Fritz, *J. Catal.*, 1999, **186**, 64.
185. F. Jayat, C. Lembacher, U. Schubert and J. A. Martens, *Appl. Catal. B*, 1999, **21**, 221.
186. W. Schießer, H. Vinek and A. Jentys, *Catal. Lett.*, 2001, **73**, 67.
187. K. Eranen, L.-E. Lindfors, F. Klingstedt and D. Y. Murzin, *J. Catal.*, 2003, **219**, 25.
188. K. A. Bethke, D. Alt and M. C. Kung, *Catal. Lett.*, 1994, **25**, 37.
189. K. Okumura, T. Kusakabe and M. Niwa, *Appl. Catal. B*, 2003, **41**, 137.
190. M. Iwamoto, H. Yahiro, N. Mizuno, W.-X. Zhang, Y. Mine, H. Furugawa and S. Kagawa, *J. Phys. Chem.*, 1992, **96**, 9360.
191. G.-D. Lei, B. J. Adelman, J. Sarkany and W. M. H. Sachtler, *Appl. Catal. B*, 1995, **5**, 245.
192. A. V. Kucherov, C. N. Montreuil, T. N. Kucherova and M. Shelef, *Catal. Lett.*, 1998, **56**, 173.

192a. K. Hadjiivanov, H. Knozinger, B. Tsyntsarskiand L. Dimitrov, *Catal. Lett.*, 1999, **62**, 35.
193. T. Ishihara, M. Kagawa, F. Hadama and Y. Takita, *J. Catal.*, 1997, **169**, 93.
194. Z. Li and M. Flytzani-Stephanopoulos, *Appl. Catal. B*, 1999, **22**, 35.
195. C. Torre-Abreu, M. F. Ribeiro, C. Henriques, F. R. Ribeiro and G. Delahay, *Catal. Lett.*, 1997, **43**, 31.
196. R. A. Grinsted, H. W. Zhen, C. Montreuil, M. J. Rokosz and M. Shelef, *Zeolites*, 1993, **13**, 602.
197. T. Tabata, M. Kokitsu, O. Okada, T. Nakayama, T. Yasumatsu and H. Sakane, *Stud. Surf. Sci. Catal.*, 1994, **88**, 409.
198. S. Matsumoto, K. Yokata, H. Doi, M. Kimura, K. Sekizawa and S. Kasahara, *Catal. Today*, 1994, **22**, 127.
199. J. Y. Yan, G.-D. Lei, W. M. H. Sachtler and H. H. Kung, *J. Catal.*, 1996, **161**, 43.
200. T. Tabata, M. Kokitsu, H. Ohtsuka, O. Okada, L. M. F. Sabatino and G. Bellussi, *Catal. Today*, 1996, **27**, 91.
201. Y. Li and J. N. Armor, *Appl. Catal. B*, 1995, **5**, L257.
202. O. Okada, T. Tabata, M. Kokitsu, H. Ohtsuka, L. M. F. Sabatino and G. Bellussi, *Appl. Surf. Sci.*, 1997, **121/122**, 267.
203. P. Budi and R. F. Howe, *Catal. Today*, 1997, **38**, 175.
204. S. Kasahara, S. Okazaki and K. Sekizawa, *European Patent* 462598 A, 1991.
205. M. Iwamoto, H. Yahiro, H. K. Shin, M. Watanabe, J. Guo, M. Konno, T. Chikahisa and T. Murayama, *Appl. Catal. B*, 1994, **5**, L1.
206. J. Y. Yan, M. C. Kung, W. M. H. Sachtler and H. H. Kung, *J. Catal.*, 1997, **172**, 178.
207. K.-I. Shimizu, A. Satsuma and T. Hattori, *Appl. Catal. B*, 2000, **25**, 239.
208. I. I. Novochinsky, P. A. Chernavsky, P. V. Ryabchenko and V. V. Lunin, *Catal. Lett.*, 1998, **51**, 191.
209. M. Haneda, Y. Kintaichi and H. Hamada, *Appl. Catal. B*, 1999, **20**, 289.
210. A. Keshavaraja, X. She and M. Flytzani-Stephanopoulos, *Appl. Catal. B*, 2000, **27**, L1.
211. S. Y. Chung, *MS Thesis, Selective Catalytic Reduction of Nitrogen Oxides by Hydrocarbons over Dealuminated Zeolite Catalysts*, Pohang University of Science and Technology, 1997.
212. C. Torre-Abreu, M. F. Ribeiro, C. Henriques and F. R. Ribeiro, *Catal. Lett.*, 1997, **43**, 25.
213. E. M. Flanigen, J. M. Bennett, R. W. Grose, J. P. Cohen, R. L. Patton, R. M. Kirchner and J. V. Smith, *Nature*, 1978, **271**, 512.
214. Z. Chajar, P. Denton, F. B. de Bernard, M. Primet and H. Praliaud, *Catal. Lett.*, 1998, **55**, 217.
215. R.L. Keiski, H. Raisanen, M. Harkonen, T. Maunula and P. Niemisto, *Catal. Today*, 1997, **38**, 175.
216. J. Y. Yan, W. M. H. Sachtler and H. H. Kung, *Catal. Today*, 1997, **33**, 279.
217. K. Masuda, K. Shinoda, T. Kato and K. Tsujimura, *Appl. Catal. B*, 1998, **15**, 29.
218. W. N. Delgass, R. L. Garten and M. Boudart, *J. Phys. Chem.*, 1969, **73**, 2970.
219. R. Joyner and M. Stockenhuber, *J. Phys. Chem. B*, 1999, **103**, 5963.
220. P. Marturano, L. Drozdova, A. Kogelbauer and P. Prins, *J. Catal.*, 2000, **192**, 236.
221. A. A. Battiston, J. H. Bitter, W. M. Heijboer, F. M. F. de Groot and D. C. Koningsberger, *J. Catal.*, 2003, **215**, 279.
222. A. V. Kucherov, J. L. Gerlock, H.-W. Jen and M. Shelef, *Zeolites*, 1995, **15**, 9.
223. H.-T. Lee and H.-K. Rhee, *Catal. Lett.*, 1999, **61**, 71.
223a. S. H. Choi, B. R. Wood, J. A. Ryder and A. T. Bell, *J. Phys. Chem.*, 2003, **107**, 11843.

224. S. Asbrink and L. J. Norrby, *Acta Crystallogr. B*, 1970, **26**, 8.
225. A. F. Wells, *Structural Inorganic Chemistry*, 5th Edn, Clarendon, Oxford, 1984.
226. H. Matsumoto and S. Tanabe, *J. Phys. Chem.*, 1990, **94**, 4207.
227. R. Piffer, H. Forster and W. Niemann, *Catal. Today*, 1991, **8**, 491.
228. H. Hamada, N. Matsubayashi, H. Shimada, Y. Kintaichi, T. Ito and A. Nishijima, *Catal. Lett.*, 1990, **5**, 189.
229. T. Nakayama, T. Yasumatsu, M. Kokitsu, T. Tabata and H. Sakane, *Jpn. J. Appl. Phys.*, 1993, **32**, 487.
230. M. H. Kim, I. -S. Nam and Y. G. Kim, *Stud. Surf. Sci. Catal.*, **111**, 1997, 213.
231. M. H. Kim, I. -S. Nam and Y. G. Kim, *J. Catal.*, **179**, 1998, 350.
232. M. H. Kim, I. -S. Nam and Y. G. Kim, *J. Chem. Soc., Chem. Commun*, 1998, 1771.
233. M. H. Kim, I. -S. Nam and Y. G. Kim, *Korean. J. Chem. Eng.*, 1999, **16**, 139.
234. L. Li and W. K. Hall, *J. Catal.*, 1991, **129**, 202.
235. E. Giamello, D. Murphy, G. Magnacca, C. Morterrra, Y. Shioya, T. Nomura and M. Anpo, *J. Catal.*, 1992, **136**, 510.
236. J. Valyon and W. K. Hall, *Catal. Lett.*, 1993, **19**, 109.
237. W. Grunert, N. W. Hayes, R. W. Joyner, E. S. Shpiro, M. R. H. Siddiqui and G. N. Baeva, *J. Phys. Chem.*, 1994, **98**, 10832.
238. S. C. Larsen, A. W. Aylor, A. T. Bell and J. A. Reimer, *J. Phys. Chem.*, 1994, **98**, 11533.
239. A. V. Kucherov, J. L. Gerlock, H.-W. Jen and M. Shelef, *J. Catal.*, 1995, **152**, 63.
240. B. Wichterlova, Z. Sobalik and A. Vondrova, *Catal. Today*, 1996, **29**, 149.
241. G. J. Millar, A. Canning, G. Rose, B. Wood, L. Trewartha and I. D. R. Mackinnon, *J. Catal.*, 1999, **183**, 169.
242. S. Matsumoto, K. Yokota, H. Doi, M. Kimura, K. Sekizawa and S. Kasahara, *Catal. Today*, 1994, **22**, 127.
243. M. Hori, A. Okumura, H. Goto and M. Horiuchi, *SAE* 972850, 1997.
244. S. Erkfeldt, M. Larsson, H. Hedblom and M. Skoglundh, *SAE* 991476, 1999.
245. S. Matsumoto, Y. Ikeda, H. Suzuki, M. Ogai and N. Miyoshi, *Appl. Catal. B*, 2000, **25**, 115.
246. N. Miyoshi, S. Matsumoto, K. Katoh, T. Tanaka, J. Harada, N. Takahashi, K. Yokota, M. Sugiura, K. Kasahara, *SAE* 950809, 1995.
247. E. Fridell, H. Persson, B. Westerberg, L. Olsson and M. Skoglundh, *Catal. Lett.*, 2000, **66**, 71.
248. G. Fornasari, F. Trifiro, A. Vaccari, F. Prinetto, G. Ghiotti and G. Centi, *Catal. Today*, 2002, **75**, 421.
249. H. L. Fang, S. C. Huang, R. C. Yu, C. Z. Wan and K. Howden, *SAE* 012889, 2002.
250. S. Geckler, D. Tomazic, V. Scholz, M. V. Whalen, D. McKinnon, J. Orban, R. A. Gorse, O. Bailey and J. C. Hoelzer, *SAE* 010510, 2001.
251. Y. Tamura, S. Kikuchi, K. Okada, K. Koga, T. Dogahara, O. Nakayama and H. Ando, *SAE* 010254, 2001.
252. J. R. Asik, D. A. Dobson and G. M. Meyer, *SAE* 011299, 2001.
253. A. Konig, G. Herding, B. Hupfeld, T. Richter and K. Weidmann, *Top. Catal.*, 2001, **16/17**, 23.
254. A. Peters, H.-J. Langer, B. Jokl, W. Muller, H. Klein and K. Ostgathe, *SAE* 980191, 1998.
255. A. W. Gray and T. W. Ryan, *SAE* 971676, 1997.
256. K. Epping, S. Aceves, R. Bechtold and J. Dec, *SAE* 011923, 2002.
257. J. Martinez-Frias, S. M. Aceves, D. Flowers, J. R. Smith and R. Dibble, *SAE* 012869, 2000.

258. M. Sjoberg, L.-O. Edling, T. Eliassen, L. Magnusson and H.-E. Angstrom, *SAE* 010106, 2002.
259. N. W. Cant and I. O. Y. Liu, *Catal. Today*, 2000, **63**, 133.
260. K. Nagashima, M. Nagata, K. Katou, K. Sod and M. Sugiyama, *SAE* 011724, 2002.
261. N. Macleod and R. M. Lambert, *Appl. Catal. B*, 2002, **35**, 269.
262. H. G. Karge, B. Wichterlova and H. K. Beyer, *J. Chem. Soc., Faraday Trans.*, 1992, **88**, 1345.
263. T. Liese and W. Grunert, *J. Catal.*, 1997, **172**, 34.

5

Catalysis of Solid Oxide Fuel Cells

BY STEVEN S. C. CHUANG

1 Introduction

Fuel cells have received considerable interests because of its potential for high power generation efficiency with low pollutant emissions. A fuel cell is an electrochemical device that allows the direct conversion of chemical energy to electrical energy.[1] Vast opportunities exist for using fuel cells at various power levels as shown in Fig. 1.

At 1-10 W (watts), fuel cells could be used as battery replacements; at 100 W to 1 kW, fuel cells could find military applications which require lightweight portable power sources for communications and weapon power; at 1 – 10 kW, fuel cells could supply power to residential buildings and serve as auxiliary power units in vehicles and trucks. At higher power levels, the solid oxide fuel cell (SOFC) could be an effective approach for the distributed power generation and the cogeneration (i.e., combined heat and power). Above 1 MW, the SOFC could be integrated with a turbine power plant to improve the overall efficiency of power generation and reduce emissions[2].

Most versions of fuel cells work best on pure hydrogen gas.[3,4] In particular, the low temperature proton exchange membrane requires the use of high purity hydrogen to avoid poisoning of the anode catalyst by carbon monoxide. In contrast, the SOFC could offer a number of significant advantages over the low temperature fuel cells: (i) fuel flexibility, (ii) high efficiency, and (iii) the use of low cost oxides and base metals as electrodes.[5,6] The present reforming technology provides the flexibility in the choice of hydrocarbon fuel ranging from nature gas, coal gas, gasoline, diesel, and jet fuels.[7] Despite these significant advantages and a number of successful demonstration projects, most SOFC developments have been hindered by the high fabrication cost.

The current approaches for decreasing the cost are reduction in operating temperature of SOFC and elimination of the external reformer by developing effective catalysts.[5,6,8,9] The low temperature operation will allow the use of a wide range of low cost metallic materials for interconnects and reduce thermal stress in the fuel cell components, resulting in improved reliability and longer cell life.[10] To address the technical barriers for cost reduction, significant efforts have been directed toward the search for O^{2-} conducting solid electrolytes as well as effective catalysts for the electrochemical oxidation of fuel and the reduction of

Catalysis, Volume 18

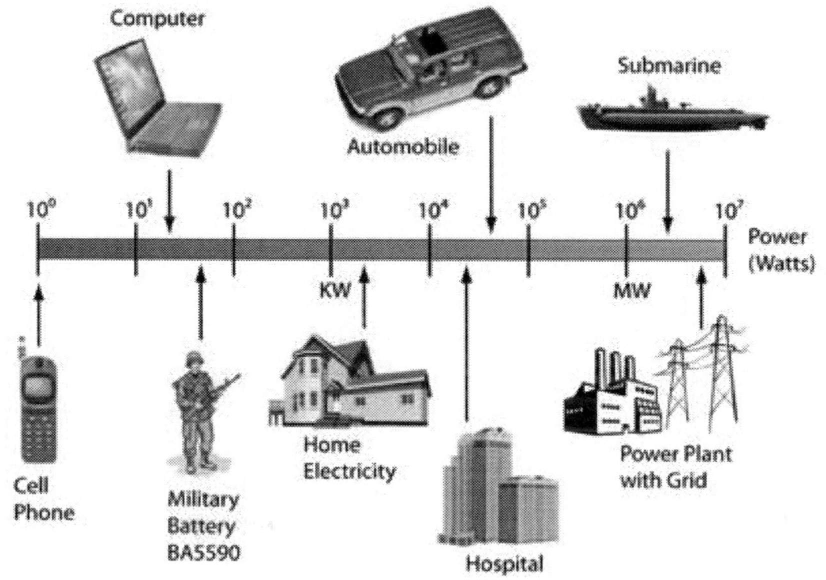

Figure 1 *Fuel cell power spectrum*

O_2 at low temperature (500 – 700°C).[5, 8, 9, 11-17]

This paper will provide an overview of SOFC catalysis and discuss the challenge as well as scientific and technological issues in the SOFC research and development.

2 Basic Principles of SOFC

Figure 2 illustrates the basic operating principles of a SOFC. The SOFC consists of an oxygen anion, O^{2-} , conducting solid oxide electrolyte dense (i.e., nonporous) membrane layered between the catalytically active anode and cathode. Typically the electrolyte is yttrium-stabilized zirconia (YSZ); the anode is Ni cermet; and the cathode is lanthanum strontium manganite (LSM). The Ni cermet is a solid mixture of Ni metal particles and YSZ particles in which Ni metal serves as an electronic conductor and an electrochemical catalyst. The Ni metal particles should be connected together so that the electrons produced form the electrochemical reaction can reach the Pt wire current collector shown in Fig. 2(b). The cathode in Fig. 2 (a) and the anode in Fig. 2(b) have the porous structure which facilitates H_2 and O_2 to access the three phase boundary (i.e., the interface among electrolyte, anode catalyst, and current collector). In the cathode side, oxygen from air adsorbs on the cathode catalyst and combines with electrons to form O^{2-}, which then diffuses through the YSZ electrolyte membrane to the anode side. The O^{2-} reaching the anode catalyst surface further reacts with the fuel molecules such as H_2 to produce H_2O. The electrochemical oxidation of each

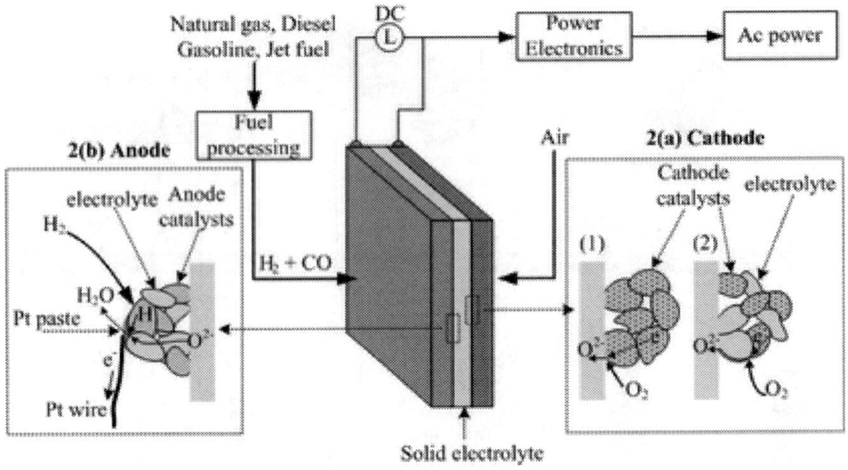

Figure 2 *Basic structure of a solid oxide fuel cell*

H_2 molecule releases two electrons with an electrical potential (i.e., voltage) of 1.055 volt at 800°C in an open circuit.

Theoretically all the combustible materials can serve as a fuel to the SOFC via the direct electrochemical oxidation on the anode. Table 1 lists a number of fuel reactions with ΔG° (i.e., the change in the Gibbs free energy) and ΔH° (the change in the enthalpy), E° (theoretically reversible potential or voltage), and the fuel cell efficiency.[4] Fuel cell efficiency is defined as[18]

$$E\ (\textit{fuel cell efficiency}) = \Delta G^\circ/\Delta H^\circ$$

For most oxidation (i.e., combustion) processes, the magnitude of ΔG° is about 90-95% of ΔH°. The fuel cell efficiency is the intrinsic maximum energy conversion efficiency that defines the upper limit of the practical efficiency. Under ideal (i.e., reversible) condition, the fuel cell achieves the maximum energy efficiency and produces the theoretically reversible cell voltage (i.e., E°_{rev}). Under practical conditions where the electrical current is drawn from the fuel cell, the polarization on the anode and the cathode as well as the ohmic resistance of the electrolyte reduces the cell voltage and cell efficiency. The majority of fuel cells are designed to operate around 70% of the intrinsic maximum energy conversion efficiency. The efficiencies of all commercial and demonstrated fuel cells operated under practical conditions are significantly higher than those of a Carnot heat engine and thermal power stations: nuclear at 30-33%, natural gas at 30-40%, coal 33-38%, and oil at 34-40%.[1, 4] It is interesting to note that the fuel cell efficiency of carbon oxidation exceeds 100% in Table 1 due to a positive change in entropy for the reaction. This reaction would require an uptake of heat from the surroundings to the fuel cell.[18]

All attempts to drive fuel cells directly using primary fuels, such as coal and oil, have not been successful due to fouling of the catalyst surface.[19] The fuel processor, shown in Figure 2, is a catalytic reformer which is used to convert a wide

Table 1 *Thermodynamics and theoretically reversible cell potential for fuel cell reactions*

Fuel	reaction	n[1]	−ΔH° (kJ/mol)			−ΔG° (kJ/mol)			E°_{rev} (Volts)[2]			Efficiency %		
			25°C	600°C	800°C	25°C	600°C	800°C	25°C	600°C	800°C	25°C	600°C	800°C
Hydrogen	$H_2 + 0.5O_2 \rightarrow H_2O$ (l)	2	286.0	244.7	246.4	237.3	214.1	203.6	1.229	1.109	1.055	82.97	87.48	82.62
Methane	$CH_4 + 2O_2 \rightarrow CO_2 + 2H_2O$ (l)	8	890.8	800.1	799.9	818.4	801.0	801.4	1.060	1.038	1.038	91.87	100.12	100.19
Propane	$C_3H_8 + 5O_2 \rightarrow 3CO_2 + 4H_2O$ (l)	20	2221.1	2039.8	2041.6	2109.3	2105.8	2127.6	1.093	1.091	1.103	94.96	103.24	104.21
Decane	$C_{10}H_{22} + 15.5O_2 \rightarrow 10CO_2 + 11H_2O$ (l)	66	6832.9	6342.1	6351.4	6590.5	6642.8	6741.8	1.102	1.043	1.059	96.65	104.74	106.15
Carbon monoxide	$CO + 0.5O_2 \rightarrow CO_2$	2	283.1	283.6	283.3	257.2	230.7	213.1	1.066	1.195	1.104	90.86	81.33	75.23
Carbon	$C + 0.5O_2 \rightarrow CO$	2	110.6	110.2	110.8	137.3	164.4	182.4	0.712	0.852	0.945	124.18	149.25	164.60
	$C + O_2 \rightarrow CO_2$	4	393.7	393.8	394.1	394.6	395.1	395.5	1.020	1.024	1.025	100.22	100.33	100.36
Methanol	$CH_3OH + 1.5O_2 \rightarrow CO_2 + 2H_2O$ (l)	6	726.6	672.9	672.6	702.5	704.24	714.78	1.214	1.216	1.235	96.68	104.66	106.28

(1) The number of electron produced per each fuel molecule or each atom in solid carbon.

range of commercial gas, liquid, and solid fuels to a fuel gas reformate suitable for the fuel cell anode reactions. The major reformate components are H_2 and CO. Due to its high reactivity, H_2 which produces a higher current density than CH_4 and other hydrocarbon fuels in the SOFC is the most effective fuel for the SOFC.

Although the basic principle and the structure of the SOFC is straight forward,[1] the design and fabrication of a high performance SOFC is a very challenging task. The difficulties arise from the high temperature operation and the lack of techniques to control the anode and cathode porous structures as well as the three phase boundary.[20] The three phase boundary is the interface where electrons in the metal wire, O^{2-} ions from the electrolyte, and fuel molecules on the catalytic oxidation site come together as shown in Fig. 2(a). Although high temperature operation allows the use of low-cost base metals and oxides as electrodes and provides high quality heat, the use of elevated temperature (above 800°C) for the SOFC operation creates the problems of searching for the appropriate materials to serve as electrolytes and electrodes. The electrolyte should possess (i) high O^{2-} conductivities and minimum electronic conductivities; (ii) good chemical stability with respect to electrodes and fuel molecules; and (iii) thermal expansion compatible with that of the electrodes. Table 2 lists the commonly used electrolytes and their conductivities.[21] Some of these elements are rare oxides; their costs can be prohibitively high for commercial applications.

Bi- and Gd-based oxides show higher ionic conductivities than YSZ. However, these oxides are susceptible to reduction under intermediate (500 – 700 °C) and high temperatures (> 800 °C), giving electronic conductivities and causing a significant loss in the electrical potential.[22] An attempt to incorporate Ca, Pr, and Fe into Gd-based oxide to reduce electronic conductivity was not successful.[23] Extensive evaluation have revealed that the SiO_2 impurities have a severe deleterious effect on the Gd-doped ceria, $Ce_{0.9}Gd_{0.1}O_2$, (GDC).[24] GDC remains one of the most promising electrolytes for the intermediate and low temperature SOFC.[25]

The anode electrode should (i) exhibit catalytic activity for electrochemical oxidation; (ii) possess electrical conductivity; and (iii) show stability in reducing

Table 2 *Conductivities of solid electrolytes*

	Electrical conductivity (S cm^{-1})	
Oxide ion conductors	*600°C*	*800°C*
$(Bi_2O_7)_{0.75}(Y_2O_3)_{0.25}$	0.0818	0.398
$BaTh_{0.9}Gd_{0.1}O_3$	0.0596	0.223
$La_{0.8}Sr_{0.2}Ga_{0.8}Mg_{0.2}O_3$	0.0316	0.178
$(Ce_2)_{0.8}(GdO_{0.5})_{0.2}$	0.0376	0.149
$(ZrO_2)_{0.9}(Sc_2O_3)_{0.1}$	0.0141	0.0841
$(ZrO_2)_{0.9}(Y_2O_3)_{0.1}$	0.00251	0.0290
$(ZrO_2)_{0.87}(CaO)_{0.13}$	0.000562	0.0140
$La_{0.7}Ca_{0.7}AlO_{2.8}$	N/A	0.00149
$(La_2O_3)_{0.95}(SrO)_{0.05}$	0.00398	0.0178
$Ca_{12}Al_{14}O_{33}$	0.00251	0.000335

environment and resistance to sulfur-poisoning. In addition, the anode should have a physical microstructure and nanostructure, shown in Figure 2(b), which lowers the resistance to the diffusion of the fuel molecule to the three phase boundary through a network of interconnected pores. A number of approaches have been developed to fabricate the porous structure of the anode using graphite pore formers.[26] Although the three phase boundary plays a critical role in governing the performance of the SOFC, no approaches have been developed to control the microstructure/nanostructure of three phase boundary.[27]

In contrast to the reducing environment of the anode, the cathode in the oxidizing environment must perform the reduction of O_2 to O^{2-} over a wide range of O_2 partial pressures. Therefore, the cathode should have a good electrical conductivity to transport the electron to the adsorbed oxygen.

3 Fuel cell performance

Figure 3 shows a typical fuel cell voltage/current characteristic. Electrical energy is produced from a fuel cell only when a current is drawn. The cell voltage decreases as the current is drawn through the external load. The decrease in the cell voltage is a result of voltage loss or overpotential (i.e., polarization) which arises primarily from activation polarization, ohmic polarization, and concentration polarization. Activation polarization is directly related to the rate of electrochemical reactions which could be improved by the proper selection of the anode catalyst and the control of its structure. Ohmic polarization results from the resistance to the diffusion of O^{2-} ions through the solid electrolyte and the resistance to the electron movement through the electrodes. Concentration polarization is a result of depletion of the fuel at the anode or air at the cathode. Concentration polarization can be alleviated by facilitating mass transfer of the reactants to the anode and cathode surfaces through control of the pore struc-

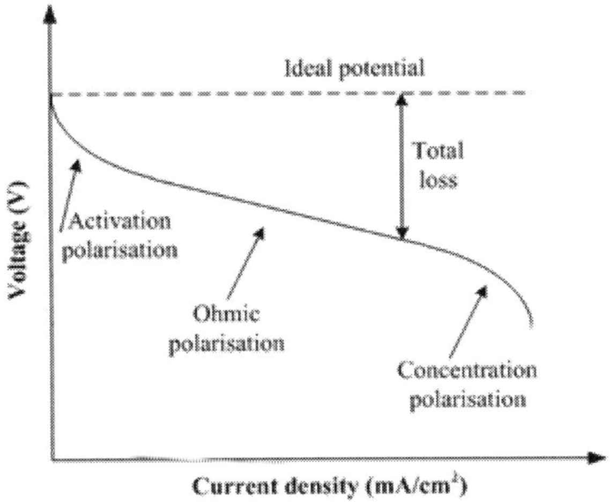

Figure 3 *Fuel cell performance curve*

ture of electrodes.

The majority of fuel cells including SOFC are limited largely by ion transport rate through membranes, i.e., ohmic polarization.[28] Studies on the SOFC with the SDC ($Ce_{0.8}Sm_{0.2}O_{1.9}$) electrolyte at 500 °C showed that decreasing the thickness of the electrolyte from 0.35 mm to 0.15 mm led to the increase in the current density from 0.2 W/cm^2 to 0.4 W/cm^2.[29] Therefore, it is highly desirable to use a thin layer of highly ionic conductive electrolyte as the membrane. In addition to the electrolyte thickness, an ideal solid electrolyte membrane has to meet the multiple requirements: O^{2-} anion conductivity, thermal expansion, stability, mechanical strength, and processing properties for fabrication of pinhole-free thin films. The electrolyte layer becomes very fragile as its thickness decreases below 0.3 mm. Therefore, the fuel cell assembly consisting of electrode/electrolyte layers has to be either externally or self-supported, as shown in Figure 4.

Ca-stabilized ZrO_2 and stainless steel have been used as an external support.[30] The key factors for the preparation of the externally supported electrode and electrolyte layers are the compatibility of thermal expansion coefficients of the substrate support and electrodes and their potential interactions. The uncontrolled interactions between electrodes and support substrates could lead to deterioration of the electrodes.

Self-supported SOFC can be classified into anode-supported and cathode-supported fuel cells. The SOFC assembly for laboratory testing has a shape of button with 1 – 2 cm in diameter and less than 500 μm in thickness. The majority of these button cells are anode-supported cells due to the easy of their fabrication as compared with that of the cathode-supported cell. These self-supported fuel cell usually possess thin (5-20 μm) electrolyte and can operate at reduced temperatures (< 800 °C). The low temperature operation is the key to decrease

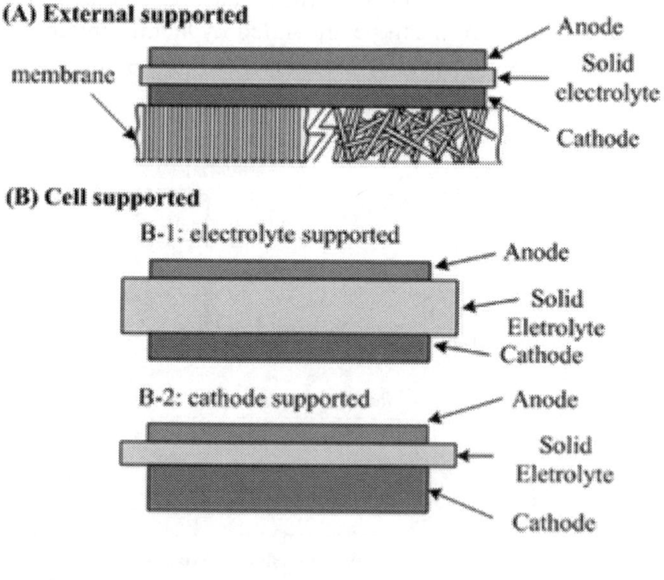

Figure 4 *Externally or self-supported cells*

the fabrication cost of SOFC. However, the low temperature operation can slow down the rate of reactions on the cathode and the anode. Thus, there is an important need for the development of effective low temperature anode and cathode catalysts.

4 Cathode

The function of the cathode is to catalyze the reduction of O_2 to O^{2-} over a wide range of O_2 partial pressure in an oxidizing environment. The cathodic materials should possess ionic and electrical conductivities which allow transport of e^- to interact with the hole produced from the reaction of adsorbed oxygen with the cathodic site. Cathodic materials can be classified into two types: electronic conducting and mixed ionic-electronic conducting. A typical electronic conducting cathode is LSM ($La_{1-x}Sr_xMnO_3$) perovskite[13]. LSM was developed from the doping of Sr into the $LaMnO_3$ perovskite structure. The doping leads to the replacement of some of La^{3+} by Sr^{2+} in the perovskite structure, creating a positively charged hole on the Mn^{3+} site, thus enhancing the electronic conductivity of the LSM cathode.

The reduction of oxygen to O^{2-} takes place at the interface of the three phase boundary. The mixed cathode which allows both electron and O^{2-} to travel along the surface and bulk extends the interface of the three phase boundary to an active zone for the reduction of O_2 to O^{2-}. The extension of the active zone depends on the oxygen diffusion, surface exchange coefficient, the microstructure and porosity of the cathode layer, shown in Fig. 2 (a).[31, 32]

The ionic conductivity of perovskite oxides is typically several orders of magnitude smaller than its electrical conductivity. The ionic conductivity of the cathode layer can be further improved by adding the ionic electrolyte materials which facilitate the transport of O^{2-} produced from the three phase boundary on the cathode to the solid electrolyte membrane. As shown in Figure 2 (a), the random distribution of perovskite and electrolyte produced a small fraction of three phase boundary, limiting the cathode performance. Approaches such as graded composite have been shown to be effective in enhancing the cathode performance.[33] One key issue in the fabrication of the cathode on the solid electrolyte film is the potential solid state interaction between cathode materials and electrolytes. It has been shown that YSZ can react with a wide range of cathode materials at 1200°C, leading to deterioration of the fuel cell performance.[13, 23] Caution must be taken in calcinations of cathode with the electrolyte to prevent the formation of a new compound at the electrode and electrolyte interface.

5 Anode

Ni-YSZ cermet anode has been widely used and studied for the H_2-SOFC.[34] The reaction on the Ni cermet can be written as

$$H_2 + O^{2-}{}_{YSZ} \underset{Ni}{\rightarrow} H_2O + 2\,e^- + V_{O,YSZ}$$

$O^{2-}{}_{YSZ}$ is an oxygen anion in the YSZ and $V_{O,YSZ}$ is an oxygen vacancy in YSZ. Due to the need for delivering the electron to the external circuit, the anode catalyst must possess both oxidation activity and electric conductivity. Ni-YSZ cermet provides a satisfactory performance at 800-1000 °C.

The main drawbacks of the Ni-YSZ anode are intolerance to the sulfur compounds as low as 2 ppm in H_2 steam and a substantial polarization loss at low temperatures (< 600 °C). The majority of the H_2 SOFCs require the use of a ZnO-based sorbent to remove H_2S in the H_2 feed stream. The alternative to Ni anode is the perovskite which possess both electric conductivity and oxidation activity at high temperature. Perovskite such as $Sr_{0.5}La_{0.4}TiO_4$ has shown excellent resistance to sulfur poisoning, however, its overall performance is a factor of 5 less than that of the Ni-cermet.[35].

The activity of Ni-cermet can be further improved by the use of different electrolytes and oxides: PrO_x. CeO_x, GDC, and SDC. PrO_x and CeO_x promote oxidation; GDC and SDC provide the ionic conductivity.[34, 36] The promotion role of these electrolytes and oxides allow the use of these Ni cermet for the low temperature (< 700 °C) SOFC.

Ni-GDC cermet not only lowered the operating temperature of SOFC but also allowed the use of CH_4 as a fuel.[30] The direct use of CH_4 or hydrocarbon provides a number of significant advantages: elimination of the use of reformer, improvement of the energy efficiency, and reduction in fabrication cost. Table 3 summarizes the performance of SOFC for the direct oxidation of hydrocarbon fuels.[14, 29, 37-41]

Table 3 *Direct oxidation of fuels in SOFCs*

Fuel	*Fuel Cell*	*Temp (C)*	*Power Density (W/cm²)*
CH_4	CeOx	900	0.07
33.3% CH_4+3% H_2	$Ce_{0.6}Gd_{0.4}O_2$	1000	0.0839
CH_4	$Cu/CeO_2/YSZ$	800	0.3
CH_4	Cu/CeO2/YSZ	700	0.1
97% CH_4+3% H_2O	Ni/YSZ/CYO	650	0.37
Butane/Sat. H_2O	Bi_2O_3-Ta_2O_5 / $Ce_{0.9}GD_{0.1}O_{1.95}$ / $Sm_{0.5}Sr_{0.5}CoO_3$	700	0.02
Ethane/Sat. H_2O	Anodeless / Pr-$BaCe_{0.8}Y_{0.2}O_{3-x}$ / $BaPr_{0.85}Y_{0.15}O_3$	750	<0.098
C_4H_{10}	CuCeO-$Sm_{0.2}Ce_{0.8}O_{1.9}$ (SDC) / SDC / LSCF	700	0.17
CH_4/O_2	Pd-Ni-SDC / SDC / $Sm_{0.5}Sr_{0.5}CoO_3$ (single chamber)	550	0.644
C_2H_6/Air	Ni-SDC/ SDC / $Sm_{0.5}Sr_{0.5}CoO_3$.(single chamber)	500	0.4
CH_4/Air	Ni-GDC / YSZ / MnO_2-LSM (single chamber)	550	0.25
C_3H_8	Ni-LSCM-GDC / GDC / LSCF	750	0.14

6 Single Chamber Fuel Cell

Lower operating temperatures allow the use of distinct catalytic properties of anode and cathode catalysts for carrying out both anodic oxidation of fuel and cathodic reduction of O_2 in a single chamber without separation of fuel and oxygen. The difference between the conventional double chamber and single chamber SOFC is shown in Fig. 5. The single chamber SOFC has no seal separating the anode and cathode, both of which are exposed to the fuel/air mixture. The single chamber fuel cell provides a number of advantages over the conventional fuel cell in (i) thermal shock resistance, (ii) rapid heating/cooling cycle, and (iii) tolerance to the crack and pinhole on the solid electrolyte membrane.

The anode catalyst should possess high electrochemical oxidation activity and minimal conventional oxidation activity; the cathode catalysts should exhibit selectivity for the reduction of O_2 to O^{2-}. Single chamber fuel cell studies using Ni-SDC as an anode and $Sm_{0.5}Sr_{0.5}CoO_3$ as a cathode showed that the anode catalyst was active for partial oxidation of C_2H_6 to produce CO and H_2 for further electrochemical oxidation on the anode catalyst surface.

$$C_2H_6 + O_2 \rightarrow 3\,H_2 + 2CO$$
$$H_2 + O^{2-} \rightarrow H_2O + 2e^-$$
$$CO + O^{2-} \rightarrow CO_2 + 2e^-$$

The cathode was highly selective for the oxygen reduction.

$$\tfrac{1}{2}\,O_2 + 2\,e^- \rightarrow O^{2-}$$

The driving force for the diffusion of O^{2-} across the solid electrolyte membrane is the difference in O_2 concentration between the anode and cathode side. This difference is a result of depletion of O_2 by partial oxidation of C_2H_6 on the anode side. Since the concentration of oxygen in the anode has a profound effect on the concentration gradient of O^{2-} across the electrolyte membrane, the flow pattern of fuel/air mixture around the fuel cell has to be carefully controlled through the flow geometry.

Table 3 shows that the power density produced from the single chamber is at the level of the conventional double chamber for the direct hydrocarbon fuel cell. Since the fuel utilization data in these studies were not provided, it remains unclear whether the overall efficiency of the single chamber fuel cell is at the level of double chamber.

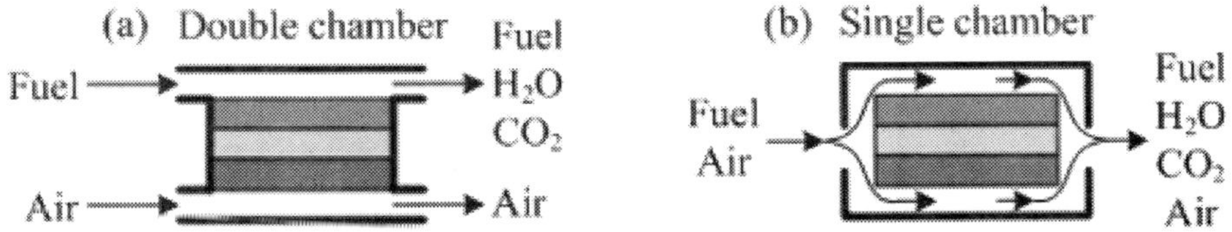

Figure 5 *(a) Conventional double chamber cell and (b) single chamber cell*

7 Conclusion

The performance of a solid oxide fuel depends not only on the compositions of the anode, electrolyte, and cathode materials but also on the microstructure of electrodes. The catalytic reactions on the anode and cathode catalysts involve the transport of electrons, holes, and O^{2-} in and out of the catalytic site with adsorbed fuel and oxygen molecules. The paths for electrons and oxygen anions are governed by the microstructure of electrodes. Although the approach for controlling the microstructure of electrodes has been a subject of extensive studies, no effective approach has been successfully developed to fabricate a three phase boundary with specific precision. The three phase boundary is generally arranged in an arbitrary pattern in the electrodes, making fabrication of SOFCs with reproducible performance a challenging task.

Extensive research has been directed toward optimizing the performance of Ni-cermet catalysts, electrolyte, and cathode, leading to development of intermediate SOFC. The key issue remained to be addressed is sulfur poisoning. Although sulfur poisoning has been a subject of extensive studies in heterogeneous catalysis, very limited work has been done on the development of sulfur-tolerant SOFC.

The direct hydrocarbon fuel cell has shown a great promise for its simplicity and significant potential for further improvement in the power density. Enhancing the current density for intermediate and low temperature solid oxide fuels requires innovation in both materials development for electrode catalysts as well as fabrication approaches.

Acknowledgements

This work was supported by the Ohio Coal development office under OCDO-OCRC3-00-4. D4.9 and the U.S. Department of Energy under DE-PS 26-03NT 41634. The author gratefully acknowledges Mr Zhig-iang Yu for assistance in preparing figures.

References

1. J. A. Fay and D. S. Colomb, 'Energy and the Environment', Oxford University Press, 2002.
2. J. Johnson, *Chemical & Engineering News*, 2004, **82**, 20.
3. E. G. Services, I. Parsons, and S. A. I. Corp, in 'Fuel Cell Handbook', ed. N. E. T. Lab, 2000.
4. A. J. Appleby and F. R. Foulkes, 'Fuel Cell Handbook', Van Nostrand Reinhold, 1989.
5. R. M. Ormerod, *Chemical Society Reviews*, 2003, **32**, 17.
6. R. J. Gorte and J. M. Vohs, *Journal of Catalysis*, 2003, **216**, 477.
7. J. R. Rostrup-Nielsen, *Physical Chemistry Chemical Physics*, 2001, **3**, 283.
8. B. C. H. Steele, A. Atkinson, J. A. Kilner, N. P. Brandon, and R. A. Rudkin, in

'Intermediate-temperature solid oxide fuel cells for power applications', Application: US
US, 2002.
9. T. Tsai, E. Perry, and S. Barnett, *Journal of the Electrochemical Society*, 1997, **144**, L130.
10. K. Kendall, N. Q. Minh, and S. C. Singhal, in 'Cell and Stack Design', ed. S. C. Singhal and K. Kendall, Oxford, 2003.
11. H. Uchida, S.-I. Arisaka, and M. Watanabe, *Journal of the Electrochemical Society*, 2002, **149**, A13.
12. T. Mori, J. Drennan, Y. Wang, J.-H. Lee, J.-G. Li, and T. Ikegami, *Journal of the Electrochemical Society*, 2003, **150**, A665.
13. J. M. Ralph, C. Rossignol, and R. Kumar, *Journal of the Electrochemical Society*, 2003, **150**, A1518.
14. T. Hibino, A. Hashimoto, M. Suzuki, and M. Sano, *Journal of the Electrochemical Society*, 2002, **149**, A1503.
15. T. Ishihara, S. Fukui, H. Nishiguchi, and Y. Takita, *Journal of the Electrochemical Society*, 2002, **149**, A823.
16. T. I. Politova and J. T. S. Irvine, *Solid State Ionics*, 2004, **168**, 153.
17. D. Perednis and L. J. Gauckler, *Solid State Ionics*, 2004, **166**, 229.
18. J. B. Jones and R. E. Dugan, 'Engineering Thermodynamics', Prentice Hall, Inc., 1996.
19. A. Hamnett and P. Christensen, in 'Electrochemical and Photoelectrochemical Energy Conversion', ed. N. Hall, Cambridge, 2000.
20. M. Mogensen and S. Skaarup, *Solid State Ionics*, 1996, **86-88**, 1151.
21. O. Yamamoto, in 'Low temperature electrolytes and catalysts', ed. W. Vielstich, H. A. Gasteiger, and A. Lamm, New York, 2003.
22. T. Takahashi, T. Esaka, and H. IWahara, *J. Appl. Electrochem.*, 1977, **7**, 299.
23. J. M. Ralph, J. A. Kilner, and B. C. H. Steele, *Materials Research Society Symposium Proceedings*, 2000, **575**, 309.
24. B. C. H. Steele, *Solid State Ionics*, 2000, **129**, 95.
25. C. Xia and M. Liu, *Solid State Ionics*, 2002, **152-153**, 423.
26. R. J. Gorte, S. Park, J. M. Vohs, and C. Wang, *Advanced Materials (Weinheim, Germany)*, 2000, **12**, 1465.
27. E. Ivers-Tiffee, A. Weber, and D. Herbstritt, *Journal of the European Ceramic Society*, 2001, **21**, 1805.
28. R. Stone and P. Szuromi, *Science (Washington, D. C.)*, 1999, **285**, 667.
29. T. Hibino, T. Hashimoto, T. Inoue, J.-I. Tokuno, S.-I. Yoshida, and M. Sano, *Science (Washington, D. C.)*, 2000, **288**, 2031.
30. S. A. Barnett, T. Lai, and J. Liu, in 'Direct hydrocarbon fuel cells', Application: US US, 2004.
31. M. Becker, A. Weber, A. C. Mueller, and E. Ivers-Tiffee, *Proceedings – Electrochemical Society*, 2003, **2003-7**, 222.
32. T. Horita, K. Yamaji, N. Sakai, H. Yokokawa, T. Kawada, and T. Kato, *Solid State Ionics*, 2000, **127**, 55.
33. N. T. Hart, N. P. Brandon, M. J. Day, and N. Lapena-Rey, *Journal of Power Sources*, 2002, **106**, 42.
34. S. P. Jiang and S. H. Chan, *Journal of Materials Science*, 2004, **39**, 4405.
35. R. Mukundan, E. L. Brosha, and F. H. Garzon, *Electrochemical and Solid-State Letters*, 2004, **7**, A5.
36. E. Perry Murray and S. A. Barnett, *Solid State Ionics*, 2001, **143**, 265.

37. B. C. H. Steele, P. H. Middleton, and R. A. Rudkin, *Solid State Ionics*, 1990, **40-41**, 388.
38. C. Lu, W. L. Worrell, R. J. Gorte, and J. M. Vohs, *Journal of the Electrochemical Society*, 2003, **150**, A354.
39. S. A. Barnett, in 'Direct hydrocarbon SOFCs', ed. W. Vielstich, H. A. Gasteiger, and A. Lamm, New York, 2003.
40. T. Hibino, A. Hashimoto, M. Yano, M. Suzuki, S.-I. Yoshida, and M. Sano, *Journal of the Electrochemical Society*, 2002, **149**, A133.
41. T. Hibino, H. Ishikawa, and S. Iio, in 'Solid electrolytic type fuel cells for generating electricity stablely at 600°C', Application: JP, 2004.